Product Innovation
Idea to Exploitation

Product Innovation
Idea to Exploitation

J.A.A. Bradbury

JOHN WILEY & SONS
CHICHESTER · NEW YORK · BRISBANE · TORONTO · SINGAPORE

Copyright © 1989 by John Wiley & Sons Ltd

All rights reserved

No part of this book may be reproduced by any means, or transmitted, or translated into a machine language without the written permission of the publisher.

Library of Congress Cataloging-in-Publication Data:

Bradbury, J.A.A.
  Product innovation.

  Bibliography: p.
  Includes index.
  1. New products. 2. Product management. I. Title.
HF5415.153.B7   1989   658.5′75   88-33848
ISBN 0 471 92169 6

British Library Cataloguing in Publication Data:

Bradbury, J.A.A.
  Product innovation: idea to exploitation
  1. New products. Management
  I. Title
  658.5′75

  ISBN 0 471 92169 6

Printed in Great Britain at The Bath Press, Avon

Contents

Acknowledgements	vii
1 The Innovation Process	1
1.1 Philosophy and experience	1
1.2 Defining innovation	6
1.3 The innovation project model	11
1.4 Product innovation phases	14
2 Key Events of Product Innovation	19
2.1 Events as innovation unit operations	19
2.2 Feasibility phase events	21
2.3 Applications phase events	28
2.4 Development and design phase events	34
2.5 Exploitation phase events	42
3 Innovation and the Innovator	46
3.1 The innovator's environment	46
3.2 Team innovators and inventor-entrepreneurs	51
3.3 Promoting innovation	56
3.4 Innovation and manufacturing industry	63
3.5 Managing innovation	67
4 Concept to Feasibility	70
4.1 The source of innovation	70
4.2 Ideas as the starting point	74
4.3 The inventive step	79
4.4 Experimentation	87
4.5 Product evaluation and definition	88

	4.6	Process evaluation	90
	4.7	Research and marketing organization interface	92
	4.8	Preparing for patent application	93
	4.9	Cost/effect considerations	94
	4.10	Project review of the feasibility phase	95
5	**Applications Research**		**97**
	5.1	Applications and secondary innovation	97
	5.2	'Properties' as product basis	101
	5.3	'Effects' as product basis	108
	5.4	Applications and market development	111
	5.5	Quality control	118
	5.6	Applications patents	120
	5.7	Project review of the applications phase	121
6	**Development and Design**		**123**
	6.1	The role of development	123
	6.2	Design and innovation	125
	6.3	Design	128
	6.4	Cost/effect comparisons	134
	6.5	Project techno-commercial assessment	136
7	**The Business Objective — Product Exploitation**		**138**
	7.1	Exploitation strategy	138
	7.2	Confidentiality, know-how and prior-art	140
	7.3	Patenting and licensing	143
	7.4	Data sheets and exploitation proposals	145
	7.5	Presentation and negotiation — the final goal!	146
8	**The Practice of Innovation**		**150**
	8.1	Introduction	150
	8.2	Getting started	153
	8.3	Experiment and feasibility	154
	8.4	Patenting strategy	155
	8.5	Weighing up the opposition	157
	8.6	Scale-up and applications	158
	8.7	Prototype evaluation	161
	8.8	Exploitation	167
References			**172**
Index			**173**

Acknowledgements

My thanks are extended in the first instance to Dr J.W. Tipping for introducing me to the publishers and for urging me to undertake this work, but I am especially grateful to Professor Derek Birchall FRS, whose encouragement, advice and help has been indispensable. F.B. Hollows and P.B. Tunnicliffe have also played an influential part in the formation of my views and experiences and this help has been supplemented by many colleagues too numerous to mention, who were kind enough to provide me with the 'space' in which to make my innovative contributions.

The contents of the book reflect the collaboration I received from several companies, i.e., ICI Plc, Steetly Berk Ltd, and Harrison Jones (Holdings) Ltd, when seeking to exploit my own projects with them. Mr Nicholls of Harrison Jones and Mr Dixon of Steetly Berk were particularly helpful.

At a personal level, my thanks are especially due to my wife for her patience during the 'writing period', when she sacrificed so much of her time to enable me to complete the work. Last, but certainly not least, my thanks go to my youngest son, Julian, whose preparation of the computer graphics has I believe greatly aided the presentation. In addition his improvements to computer hardware and software has facilitated the production of the book and ensured that a 'back-up copy' was always available to rescue me, when I pressed the wrong key — as so often I did!

1 The Innovation Process

1.1 PHILOSOPHY AND EXPERIENCE

In attempting to propound a philosophy of product innovation, I have used as a basis for study, experience gained from participation in a variety of new venture type projects, with the inevitable result that my approach to the subject reflects not only my own innovative efforts, but those of many of my colleagues who shared this field of endeavour with me. From this corporate experience I have attempted to find some kind of consistent pattern, common to all of the projects examined, which will be descriptive of the innovative process itself and which will in its broadest sense, embrace the generation of ideas, the translation of them into prototype products and their subsequent exploitation as new business ventures.

Personal experience also features in my choice of those projects which can best serve to illustrate the many different aspects of particular events implicit in the transition from 'idea to exploitation'. From the analysis of a relatively small number of such projects, each differing in kind, the key events which fall into natural groupings or phases have been identified. On ordering these in a logical way, a pattern emerges which I find is also evident in other projects which have distinctively different 'end products'. From these observations it seems reasonable to expect that a 'model' can be constructed to interpret the innovator's role at the operational level and describe the detailed interactions that constitute the innovative function in new venture projects in general.

A model of the kind described, i.e. an innovation project model, can be of value in attempting to analyse the way in which innovation has taken place in a project, i.e. it can provide a structure for an exposition of 'case histories' of individual projects. Conversely it can prove to be of particular value in the evolution of a project target and its realization, in response to a general objective, i.e. it may serve as a guide in the process of innovation. One of the more immediate advantages of such a model is that it provides a logical framework in which to explore creative events and this facilitates my attempt to convey how, when and why, innovation arises in the succeeding pages of this book.

My emphasis on the logical aspects of innovation may seem at first sight to be contradictory to the more obviously held idea of innovation as a form of creativity. My argument is that the logical structuring of the innovation model concept provides the essential complement to the creative element of innovation that finds expression 'within' its constituent events and their interactivity with each other, i.e. both creativity and logic, contribute to the realization of a new venture project and characterize the innovation project model referred to, as I trust subsequent argument will endorse.

In terms of explaining how innovation takes place, a further advantage to be gained from drawing on personal experience, is to be found in the facility afforded for identification of details, often obscured to the external observer, which explain the origins of key innovative or inventive steps. It is the innovator responsible for taking these steps, who is in the relatively unique position of being able to define clearly the content and sequence of the events which encompass them. Such information is a valuable source of reference in establishing the philosophical points to be made with respect to the *modus operandi* of the innovation process.

The above comments are not to argue that the innovator is the know-all, on the contrary, it is to expose the unpredictable nature of the way in which the innovator makes his response to an idea for a new product, which he may have originated or which he receives from others. When this insight is allied to the facility afforded by the model for logical analysis it should provide us with a picture of the evolutionary characteristic of the innovative process.

Paradoxically, while the identification and location of the creative contribution is achieved by such analysis, the creative element arises apparently without any logical thought at all! That is, it does not draw upon logic for its inception, yet always contains the inventive elements of the project. Hence the argument for creativity and logicality as the corner-stones of innovation.

In the light of these opening remarks, the case for producing a book on innovation from the inventor's viewpoint seems to be justified, but I confess that the thought of attempting to put my own ideas and experiences in writing never occurred to me until the publishers approached my employers of the time, to explore the possibility of encouraging work on a publication which would open up ideas on the subject of 'new ventures' research.

The task of responding to the invitation eventually fell to me, whereon my first thought was that it would be wise to avoid an over-preoccupation with the scientific and technological aspects of a specific new venture in the form of a long project case history, preferring instead a broader view of the subject in the form of a philosophy which would consider not only invention, experiment and illustrative examples from case histories, but would include product applications and marketing as well as the very important areas of patenting and eventual product exploitation. I hope that the result is pertinent to individual interests and meets a variety of needs.

My own limited experience in a number of widely differing 'new venture' projects has encompassed management, design, research and marketing functions, much of it within the conventional structure of a research and development department of ICI. This provided me with a basis for incorporating team innovation in my considerations, but I deferred making a start on this work until I had subsequently acquired some further and complementary experience as a freelance consultant on R & D problems involving new materials and, as events transpired, as a lone inventor-entrepreneur. Broad as this experience may appear to be, I hastily refer readers with a particular interest in the business end of the new venture project to 'works' by Drucker. These are of particular value to the entrepreneur.

The approach I have adopted, i.e. of considering innovation from the standpoint of the team innovator and the lone inventor-entrepreneur is tantamount to, 'looking at both sides of the coin' and this provides a comprehensive idea of the amazing diversity and fascination of the innovator's potential role. In this way, I hope that the substance and value of my contribution to an understanding of the subject will be improved, especially as it will also reveal among the anecdotes, more about failures of attempts to implement ideas, than successes, which I consider to be of more value to the discerning than the usual parade of case history success stories. A feature of this approach is that it also spells out the unmistakable differences in problems encountered by the corporate team innovator when compared with those of the inventor-entrepreneur.

Rationalization of my thoughts on the subject during the preparation of the 'book plan' and the encouragement I received from a number of colleagues, whose opinions I greatly respected and with whom I had spent many happy hours in the laboratory and 'market places' on new venture projects around the world, finally influenced me to get down to the task of showing what is involved in the innovation of new products, from the standpoint of an innovator and inventor, in both corporate research team and in the role of lone inventor-entrepreneur. What follows is drawn from examples found in those projects which utilize manufacturing processes for the realization of their 'product' and more specifically, those which are of a new materials category, as most of my experience of innovation derives from work in this field.

As I entered more deeply into the book planning stage it soon became clear that the subject was more complex than I first imagined, or put succinctly, 'I might have bitten off more than I could chew'! It now began to dawn upon me why contemporary publications of the type contemplated were not very evident. In spite of this I have pressed on, for I am persuaded that the role of the innovator or more specifically, the inventor, needs to be publicized in order that it is better understood with reference to the operational details and philosophy entailed. In this way I believe that what is presented here might contribute, at least in some small degree, to the emergence of a new appreciation of his role and assist in the identification, education and encouragement of those with aptitudes for innovation, perhaps under a concerted effort of government, industry and educational establishments.

A glance at the contents page will immediately show the reader that new product innovation involves a multiplicity of activities. I have concentrated upon those which are critical to the exploitation of the initial idea, i.e., in terms of achieving its successful translation into a prototype product on which a business study can be made with a view to its production and marketing — or its early exploitation through the sale of know-how and/or patent rights etc. In other words — secure resources for manufacture and sale of the product — or sell the ideas which demonstrate the project's techno/commercial potential.

How ideas emerge; are then translated into invention and progressed through the various stages of innovation up to the exploitation point, constitutes a framework in which the innovator finds his role and to which the reader may be expected to relate with some aspect of his own experience or knowledge, but I have observed that much confusion exists in the minds of many, including myself when first addressing the definition of what we mean by innovation. In the early stages of this book I have sought to clarify matters by differentiating between those activities which make up 'innovation' and in so doing lay down the basis of a philosophy which finds expression in a surprisingly disciplined methodology.

From these initial comments I hope those who are contemplating any entrepreneurial incursion into the innovation field with ideas of their own, will appreciate the duality of the task they contemplate in terms of creativity and methodology. These represent in many ways the opposite ends of the spectrum of mental activity and indicate that a disciplined approach is demanded if the creative individual is to carry his ideas forward in a way likely to win support from those whose assessment of any proposal is made on an objective business orientated basis.

For the sake of emphasis — innovation finds its origins in creativity, but its realization in a logical interpretation and development of the substance of creativity, or as a popular song puts it — you can't have one without the other!

In its broadest sense, the process of achieving the transition from idea to exploitation, characterizes product innovation. The objective here, is to contribute to the understanding of the innovation process by providing a participant's viewpoint of how this transition occurs, which I hope will be of interest and use to a wide variety of practitioners, students, entrepreneurs, and especially those already interacting with researchers and inventors in the pursuit of new products. In particular I hope it will be found of value to any who recognize a need for a greater appreciation of how the innovator and inventor define and perform their roles and who wish to use this knowledge as a basis for a critical reappraisal of their own relationship to them, as well as to others involved in the generation of new business opportunities for manufacturing industry.

In presenting these views of innovation I have kept in mind the interests and needs of those who have no experience or contact with research and manufacturing, but who find that they have an 'idea' for what can be defined as a new product, but little knowledge of how to implement it. The alternative title for this book, i.e., *Idea to Exploitation* is meant to indicate to those looking for practical guidance in 'how' to achieve the exploitation of their idea, that they may expect to find some help in these pages.

One important group of readers to whom I address my remarks, are the science students who are still searching for the field of application in which to apply their knowledge and other capabilities. It is this group that might find the experiences narrated indicate to them that their special aptitudes lie in that unusual combination of scientific understanding and creativity which will find its fullest expression as participants in the innovation process. If this leads them into entering the research field with an eye to technological exploitation of ideas, or to embarking eventually on an inventor-entrepreneurial role I will feel that the task has been worthwhile, especially if the result of their endeavour yields the new products which can be a springboard for new business opportunities and provide of course, a personal triumph for themselves in terms of fulfilment and financial reward!

To complement the above, those who have a creative flair but lack the professional training and education to fully exploit it, might take heart by recognizing that if they are to fulfil their own aspirations as innovators and inventors, they would be wise to get down to some form of study aspects of the professional field in which they find most of their 'ideas' for new products.

Having made the appeal to emergent innovators, I must add an early warning to those who think that the invitation to take on the innovator's role is an easy and tempting outlet for their energies. The innovator's life is often fraught with misunderstanding, disillusionment and disappointments as well as with the flush of success. Not least is the fact that he has often to fight a lone corner in pursuing the kind of career in which he can achieve a full realization of his own potential, a remark I consider applies regardless of the professional discipline in which the innovator may be trained or specialize in.

The positive point I wish to stress with regard to those searching for the field in which to concentrate their efforts is, that if they identify in themselves a creative drive and ability, they are only likely to find real fulfilment in a task, such as innovation, which enables them to use such skills and aptitudes. I say this with some feeling and conviction since I have experienced the trauma of finding oneself on the outside of a new venture project while bursting with ideas and interest in aiding its realization. On eventually breaking into the research field as an applications specialist, a role that is essentially one of the innovative application of new products and their development, I knew that I had found the job that allowed me the most satisfaction. I trust my experience may be of benefit in helping others to make their career choices and that it may serve to demonstrate that innovators find that their career fulfilment is very much a thing of their own making.

In the limit, creativity cannot be contained. It makes its own opportunities and creates its own environment whenever it instinctively finds a door is open, or one that can be opened.

In drawing these opening remarks to a conclusion I particularly hope that my efforts and style of presentation, will make this book readable, enjoyable and illuminating to that much wider lay public who have a thirst for information on the developments which take place in the field of science and technology and who may be intrigued to know exactly what innovation is and how invention and inventors play their part in the generation of new things.

1.2 DEFINING INNOVATION

On seeking to determine the public's conception of what innovation is, a wide variety of reactions are encountered which range from a stare of incomprehension, to attempts to simply define it in one or two words. Typical of the latter are: new ideas, invention, design, development, etc. Such descriptions are subscribed to by some of the practitioners of innovation themselves, but this should not surprise us as they see its function concentrated into their own specialist sphere of activity. The fact is that innovation can embrace all of the descriptions given, but individually and collectively such a list of definitions still fails to convey adequately its dynamic character, as I hope new readers in this subject will eventually discover.

We start our exploration with an attempt to clear up some of the misconceptions about innovation by reference to various ways of looking at its relationship to manufactured products in particular. The idea of an innovation project model, touched on earlier is then developed to give a practitioner's view of innovation as a process which links together numerous activities or 'events' whose ultimate objectives are the creation of a new product. The 'way' in which these activities interact in the form of an innovative chain forms the basis for the succeeding chapters and is central to the philosophy advanced.

Innovation is concerned with change. In its broadest sense it is therefore, the implementation of 'ideas' to effect that change. Such 'ideas' may arise from marketing and technological knowledge or from experiment. Some emerge that appear to owe nothing at all to these sources, as is evident from the diversity of innovative projects that are encountered. These range from activities as varied as introducing a new way of organizing a charity collection, to devising a new process for producing a particular 'effect' in a nominated situation or circumstance. The application which is of particular interest here, is where innovation is the realization of 'new products' which preferably possess an element of novelty in their concept and/or processing and are suitable for production by manufacturing industry. The operative words here are: ideas, novelty and manufactured products.

The generation of 'ideas' for such new products and their realization as business opportunities presents the innovator with a challenge which may embrace such diverse activities as market appraisal and design, as well as many aspects of research and development. This demands considerable and persistent effort in terms of creativity and methodology from the innovator as already intimated, which may appear to present us with a paradox, but on exploring more fully the nature of innovation it will be seen that the two aspects are in fact complementary in operation.

While 'ideas' are originated or formulated, by an individual, including those emerging from 'collective idea seeking methods' such as corporate brain-storm sessions, their translation into a 'prototype product' and their subsequent exploitation in business opportunity terms, requires the input of many people with widely differing skills and aptitudes. Such resources often exist within the environment in which the ideas have been generated, usually in the form of a research project team, backed up with marketing and business organization personnel, in the case of truly corporate enterprise.

Where ideas emanate from a lone inventor-entrepreneur a great deal of improvisation is often called for, owing in general to limited resources in terms of manpower and knowledge, but in both categories of innovation, many projects cannot be advanced without the aid of external collaborators, often identifiable among potential customers for the product, or among suppliers of feedstock materials or components, etc. Obviously a variety of research institutions and universities also form an important part of the spectrum of collaborative support available to innovators in general.

From these initial observations it is evident that the successful realization of an 'idea' for a new product will depend markedly on the corporate endeavour and will, of those participating in the new venture environment, but most of all it requires from the innovator, a highly personal commitment to communicate and collaborate with all participants in the project, particularly at the inception stages. This implies his sustained belief in the project's feasibility and demands that he should be able to build up confidence in it in the eyes of participants and potential purchasers or investors.

To achieve all this, it is evident that the innovator will need to be gifted with plenty of energy to drive the 'idea' through a systematic programme, often entailing a rigorous period of experiment and testing, which in some cases yields invention, invention being distinguished from innovation, an issue which is elaborated upon later. Above all else, the innovator must have the ability to respond to the inevitable problems to be encountered in the project with a creative and receptive capacity for change in his own 'ideas'.

The innovator is the agent of change. Technological change by its nature precipitates many problems affecting the techno-commercial aspects of a project, but the consequences of 'change' also greatly affects individuals, among whom the innovator and his colleagues are no exception. It is not surprising therefore that the innovator and the inventor in particular often encounter a measure of frustration in 'new venture projects' by virtue of the resistance to 'change' which is inherent in us all to a varying degree. These changes are the expression of the dynamic nature of the innovative process involving as it does, a multiplicity of interactions which inevitably pass through the field of activity of numerous participants in the project, the majority of which impact upon the innovator or inventor to a surprising degree.

It follows, that getting an 'idea' for a new product across to others, demands more than a technical exposition of how it may be interpreted. The human factor asserts itself strongly since these interactions bring the innovator into close working relationships with other participants in the project and occasionally into conflict with them! Such relationships extend in a similar fashion to external collaborators. Happily the majority of the interactions in which the innovator is involved are usually of an extremely amicable and constructive nature, no doubt due to the participants sharing in the common objective of achieving a successful 'end product', especially where this is coupled with the inherent satisfaction and stimulus to be found in the exchanges of ideas, knowledge and skills within the new venture project team and/or its collaborators. It is this environment of continuous exchange and interaction that creates the innovative project.

I do not think it is going too far, to claim that the innovative activity described, provides a sense of excitement for the creativity inclined which finds its parallel in the visual artist's vocabulary. Even the dual demands of creativity and logical progressive development work, which is a feature of the product innovation process, finds analogy with the spontaneous impulse and complementary intellectual response of continuous visual refinement, employed by the artist-painter, when he makes his first mark on a canvas and then succeeds it with a combination of both creative action and analytical assessments of what has already been stated.

Clearly the initial 'idea' or impulse is of paramount importance, regardless of its origin, but its eventual final expression depends on the response made to it, in the form of its further expansion and interpretation.

One of the key features of our theme, relevant to professional and layman alike, which casts much light on what innovation is about, is an analysis of the 'sequence of events' which constitutes the innovative process structure itself. This reveals how, when and where innovation emerges and in a similar way, how, when and where invention can arise. This approach is used in the succeeding chapters where a number of practical examples of new venture projects are analysed, drawn from a number of fields, each of which highlight the wide diversity of influences that shape such 'events' within a common framework.

As the theme is developed, the emphasis on the 'sequence of events' will also be seen to substantiate the claim that innovation is a dynamic process, involving the recycling of both ideas and information in pursuit of objectives that themselves often undergo several changes before the project reaches the stage where its final marketable 'product' can be identified. This is in much the same way that the artist continually redefines his subject, until it reaches the stage where he recognizes it as a 'finished product'.

To extend the analogy a little further, it is said that the artist's main problem once he has evolved his initial idea is 'when to stop'. Or put another way, 'what to leave out'. The innovation-technologist is in much the same plight when he suddenly finds that he is developing a plethora of ideas which begin to swamp his ability and resources for investigation and which may eventually detract him from the most effective interpretation of the initial creative concept.

Even in the case where innovation is initially targeted on a specific product during the course of a project, 'ideas' can subsequently change, as a result of new information which makes it evident that a redefinition of objectives is required. This does not imply indecisiveness or inefficiency since by its dynamic nature, the innovative process precipitates such changes.

Using again the analogy of the 'visual artist', in the creation of a work, the initial statement on the canvas is often destroyed by frequent retranslation of the initial idea into a new image. A classic illustration of this was seen in a film showing Picasso at work on a drawing whose subject started life as a 'hen', but ended its creative period as an entirely different subject — Yes! you guessed correctly; it did transform into a woman's head — as a consequence of continuous redefinition! You may not always share the appreciation held by others of such a controversial artist, but what was demonstrated was the creative act at work. Its analogy to the innovation of the most technological of products is evident and provides a graphic illustration of 'how innovation is achieved', i.e. innovation is a reiterative process! What we start out to achieve is so often fundamentally different from what we eventually realize in a new venture project.

In the light of the above remarks the following comments may assume clearer significance for us, i.e., when a project is commenced on the basis of a general objective, instead of a sharply defined product target, the innovation process explores the objective from a variety of directions, in the course of which completely different product possibilities emerge and as a result, product targets are identified within the context of the initial objectives. Priorities are then selected from the options which present themselves. By contrast, where design forms a significant part of a project it can be 'targeted against a specification' and hence clear design objectives can be set at the commencement of it. In the latter case the opportunity for invention within the innovation process may be limited by the strictures imposed on the project, but innovation through design is itself a subject for our consideration later, as the chapter on Development and Design reveals.

The route by which progress is made towards 'product exploitation', is seen from the foregoing, to be strongly dependent upon the responsiveness of the innovator to the changing 'pattern of events' as the initial idea matures often through invention and development. This emphasis on individualism in the innovative process is inevitable, but the input of a project team as a whole is demonstrably a major and complementary factor in attaining the exploitation goal.

Looking at the place of innovation in the priorities of manufacturing industry, innovation has received much publicity in the United Kingdom in terms of what is needed for national economic growth, yet as my opening remarks state, ambiguity still exists in some quarters about what innovation actually consists of in relation to new manufactured products. Before developing my own interpretation in this respect I wish to introduce the reader to the work of the Design Council in the UK, who have recognized the need for greater awareness, at all levels of society, of the role of innovation in relation to manufactured products. This has resulted in their inception of a consultancy service aiming to provide information on various innovative aspects of a range of manufacturing projects in the form of 'case histories' supported by a small exhibition of 'products' at their London Design Centre.

The Design Council have also published a pamphlet which examines 'Innovation in relation to Manufactured Products'. In it we read that, 'Innovation in relation to manufactured products involves all the steps from research, through development, prototype construction and manufacturing, to selling in the market' (Gardiner and Rothwell, 1985). The authors illustrate their work with several models of the innovation process; see Figs. 1 and 2.

Fig. 1 shows two traditional views of the process, both of which follow a linear path, i.e., the first of which is the 'Technology Push Model', commencing with basic science, then through applied science and manufacturing to the market. The second view is complementary, i.e., the 'Market Pull Model', beginning with market need and responding with development and manufacturing, concluding with sales.

By contrast, Fig. 2 presents an 'interactive model' which recognizes that the majority of innovative processes in practice, rarely follow either the 'market pull' or 'technology push' models. Instead they take a non-linear route, where need and technology continuously interact. As Gardiner and Rothwell describe, 'the innovative process may start with market need or new technological capability but whichever is the case is not critical to the outcome, as long as the innovating company forms the interactive link between the two in its development'.

Drucker (1985), in his *Innovation and Entrepreneurship* provides us with an excellent survey of the range of innovative sources as well as setting out principles of innovation, which provides us with a range of do's and don'ts for its successful practice. His description of 'new knowledge' as one of the sources of innovation is most relevant to the field I wish to explore.

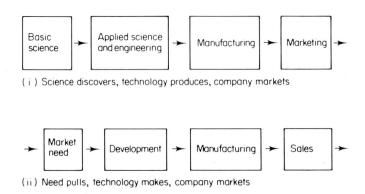

Figure 1 The traditional view of innovation (Gardiner and Rothwell, 1985) (Reproduced by permission)

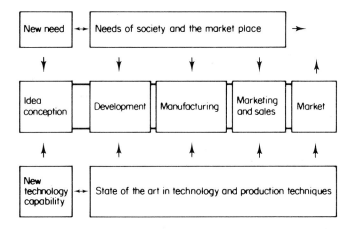

Figure 2 Interactive model of innovation (Gardiner and Rothwell, 1985) (Reproduced by permission)

It might be asked, what need is there for expansion of these ideas on innovation? My reaction is, that at both practitioner and management levels we meet an ambiguity of understanding in the subject and while aware of my own limited abilities in the field I am persuaded that a philosophy directed at explaining the 'how' of innovation demands a model of its functionality which interprets the 'operational level' as intimated earlier. The next section introduces an attempt to construct such a model.

1.3 THE INNOVATION PROJECT MODEL

New venture type projects can be subdivided into several phases of activity, each of which encompasses a range of distinctive events. The relationship of both phases and events is often highly interactive, i.e., they effect the development of each other as a result of the influence of new information or decisions which they each contribute to the project as a whole.

This characteristic of the new venture project shares common ground with the 'interactive model' of Gardiner and Rothwell (Fig. 2), who identify the broad interactions between 'market needs' and 'technological state of the art' as the implementation of an innovative project. Where the proposed innovation project model differs from this is in the scale of events at which the interactivity is identified, i.e. it is applied in the 'practitioners' sphere of activity involving those interactions which occur at a more intimate and detailed level of a project and predominantly between the distinctive events that constitute the building blocks of the innovative process itself where the innovator and more specifically the inventor have a key role to play.

On applying this latter view of the innovative process, to the field of 'new venture' projects targeted on the creation, application and exploitation of products such as 'new materials', it is evident that a recurring pattern of events

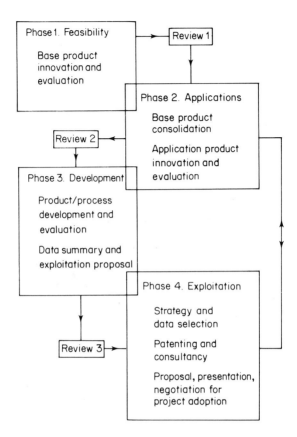

Figure 3 The Innovation Project Model — idea to exploitation. (The four phases are amplified in Figs. 4, 5, 6 and 7)

takes place as suggested in Section 1.1. This involves interactions with other events of an allied nature as well as with those concerned with complementary roles. On inspection, it becomes obvious that the pattern of events is not a simple linear progression and that any model of innovation needs to be expressed in the form of a 'flow chart' which displays both transverse interactions between adjacent events as well as a lateral or sequential progression of events.

I have sought to provide a broad outline of this way of viewing the innovative process in the form of the Innovation Project Model (Fig. 3), which seeks to interpret the interactions indicated and is designed to show clearly, the relationship between the various phases in particular. In constructing the model I have used the Venn diagram form to indicate the transverse interactions in the simple relationship between adjacent phases which are discussed in detail in Section 1.4. The model is subsequently amplified in Chapter 2 where it is expanded to show the internal pattern of events that constitute each phase. The transverse interactions between events mirror those of the phases as the relevant figures referred to in Chapter 2 show.

From the foregoing it will be evident that Chapters 1 and 2 provide complementary views of two distinct facets of the Innovation Project Model, i.e., one is the pattern of the phases and the other the pattern of events constituting the phases. When considered together they form the comprehensive picture provided by the Innovation Project Model (Fig. 3).

Progression from idea to implementation and eventual exploitation is portrayed as the innovation objective, but this may take several forms. The prime options are the leasing or sale of innovative know-how or patent property or the seeking of sanction to invest capital in the design and construction of production plant with a view to establishing a commercially viable new venture business based on a new product using internal or external resources.

The inception of innovation is subsequently shown in Chapter 4 to be centred on three critical events, i.e., idea, invention and experiment. These operate as single events or pairs of events, or by all three 'events' acting at times synergistically. Analysis suggests that the permutations of these specific events, in terms of which one is dominant in the 'set', also affects the way in which Innovation is initiated.

Interactions are the *modus operandi* of the innovative process where further ideas and sometimes secondary invention occurs. One feature of these interactions is to knit together the events, i.e., provide coherence to the innovative process by ensuring that the project continually responds to both external and internal sources of information and experimental results, often by generating secondary ideas and invention.

The secondary ideas and secondary invention referred to, often arise as a result of interactiveness between applications research and the base product. In this context it forms an important part of the model whose dynamic characteristics illustrate the point made by Gardiner and Rothwell that 'innovation is a process containing many false starts and rethinks', i.e. 'a reiterative process'. The analogy already given of the visual-artist and his process of destroying and recreating is again relevant.

As a consequence of the recycling of information, the introduction of secondary invention may occur at any radom point of the innovation chain of a project. By what I imagine is a horizontal or transverse thought process some 'event' is matched with other information or 'event' to generate ideas from which invention often springs. The analogy to the Gardiner and Rothwell interactive model is clear, but the interactions occur at the personal level of involvement for the innovator in the innovation project model. As a result of this horizontal or transverse thought process or 'coupling' we sometimes have the phenomenon of an innovative project precipitating at some random stage the start of a new innovative chain in which is no longer related to the initial idea of the parent project.

Understanding how these interactions occur, enables us to follow up innovative creativity with logical organization and interpretation of the subsequent growth of the innovation chain. We are therefore likely to improve our own innovative performance and develop an effective management of the innovatively based project through a heightened perception of innovative and sometimes inventive events, as they arise.

The reader will have observed, that I have again emphasized the conjunction of creativity and logicality. The point I wish to stress is that innovation and invention in particular are the result of intuitive ideas which do not resort to logic in their creation, but following the idea generation, logical ways of translating the idea into a viable product and business proposition is often introduced. *Innovation is therefore a creatively initiated process which is then developed and progressed to a definable goal by the application of further creativity allied to logical analysis and work organization in which the creative element continually introduces 'change' as a 'horizontal shift' in the logical progression of the chain.*

It seems to me that the 'project model' of the innovation process can be tailored to fit a range of different types of project, but with the condition that its structure incorporates the four innovation phases which I identify as:

(1) Feasibility.
(2) Applications.
(3) Development and design.
(4) Exploitation.

Reference to the model should assist in the management of a variety of new venture type projects, serving as both guide and check on progress along the innovation chain of events, terminating at the critical business decision point when the project reaches the exploitation stage.

In identifying the key events in the process it has been important to refer to project case histories to illustrate the interactiveness and 'shift' in direction already referred to as, typical of new venture projects. The examples I have selected are almost exclusively those concerned with the innovation of 'new materials' in which I have played some small part and as a result, this gives the description of the events involved a particular emphasis towards those well understood operations that must be carried out in the research and development of them. In spite of this emphasis I consider that the logic of the innovation chain will be found to be applicable to a wide variety of projects simply by redefining some of the events that make up the various phases.

1.4 PRODUCT INNOVATION PHASES

We have already seen that the process of innovation with respect to manufactured products as illustrated by the model (Fig. 3), breaks down into four phases. In the first three of these, the phases culminate in a major project review, but in the case of the exploitation phase it concludes with the negotiation for the transfer of the project to a new business venture status, in the form of capital sanction for a small-scale production plant — or alternatively as the sale of know-how and/or patent rights to others to implement the transition phase from development to some form of commercial production.

Fig. 3 shows the four phases overlapping each other, implying that while the logical progression through the innovation chain is primarily sequential, adjacent phases may be entered before completion of the 'contemporary phase' as information is exchanged or interactions take place. In the model, the areas of overlap indicate the interactive field which is also supplemented by feedback loops where information is realized discretely and then recycled. As a particular example of this, note should be made of the exploitation phase influence on the applications phase and the development and design phase in both of which, exploitation is part of the strategy of the innovative process. It is towards the conclusion of the project that the exploitation phase asserts itself fully as the focus of negotiation of a successful business outcome for the project.

The broad content of the four phases of the model is now summarized as follows:

Phase 1 — Feasibility

This is concerned with showing how initial ideas for new products are generated and their feasibility is determined. It concentrates on the interaction between idea, experiment, invention and evaluation, which leads to definition of the basic product. The phase culminates in a major project review, at which point, the status of the project in terms of invention and feasibility is examined and the decision is made for continuing the base product evolution and starting the applications phase.

An applications strategy must be formulated at this review stage, which will cater for the various ways of organizing the next period of research and the best way of exploiting the potential commercial value of the invention, by seeking ways of utilizing the base product or adaptions of it in various applications as indicated in the first instance by its properties. This signifies the transition from the feasibility phase to the applications phase through applications research.

Phase 2 — Applications

The phase combines the dual roles of base product consolidation and applications innovation, i.e., it shows the way in which applications research takes the initial base product, following the feasibility phase project review, and starts a secondary chain of innovation activity that mirrors in most respects that employed for the base product. Its sequel is in the innovation of secondary products for specific applications based on utilizing the basic product as the starting material, feedstock, or component, whenever possible.

Also in this phase, the properties, effects, processability and concept of the base product will be consolidated in relation to the primary application with which it is already associated.

At the conclusion of the applications phase, data on the comprehensive range of applications or secondary products backed by market research and patenting, forms the basis for the applications phase project review, when the decision to proceed to the development phase is taken, i.e., with respect to both the base product and secondary applications products. A development strategy is worked out at this stage to ensure that the initial base product and any secondary or applications products are advanced to a stage where they can satisfy market criteria of acceptability.

Preliminary consideration must also be given at the applications phase review to a strategy of selling-off the know-how and patent property instead of developing the project in-house any further, but action in this direction may be considered to be premature. In the latter case the issue will be reopened on reaching the development phase project review and carried through into the exploitation phase itself, unless the exploitation scene changes!

Exploitation of the initial invention as early as possible is particularly relevant to the lone inventor-entrepreneur, but this option will also feature in the considerations of the corporate research project team.

Phase 3 — Development and Design

This phase advances the project from the applications phase project review through the prototype product laboratory stage of development and design while initiating further market and applications research, which when combined with the research economics business and cost studies, provides all the information needed to initiate the concluding exploitation phase. It is on this basis that the commercial future of the project will be assessed. The introduction of the prototype for production and market evaluation of course introduces us to the important part played by project and process design.

The product design amounts to developing changes in the laboratory 'product' to improve or optimize its effectiveness and to increase its propensity for commercial scale production by economic processes, but in particular it will ensure that the product can meet the test criteria and specifications which will be increasingly encountered at this stage. The design of a small-scale semi-technical plant is often closely associated with the prototype product design. The combination of these two design events, supplemented by an adequate quality control system, facilitates the sampling of the market and verification of data for full-scale plant process design.

Once prototype products can be produced under satisfactory quality control the development phase can pick up the applications phase collaborative evaluation programmes, undertaken with potential customers and feedstock suppliers, among others. In these cases the prototypes can receive an evaluation which will extend the value of the data already acquired under the application phase when using only laboratory samples. The outcome is a full reappraisal of the product based on a facsimile of the expected full-scale product, plus the added value of some modelling of quality control of the processing as well as the product.

Perhaps the important contribution made by the development phase will be recognized in the completion of the product data sheet which incorporates all of the key information on the project, gained during the feasibility and applications phases, but transformed by the performance data gained from the development and evaluation of the prototype product against specifications and standards dictated by legislation and user requirements.

This data forms the basis for the development project review where it serves in the assessment of the project's achievements in realizing product and process. When the technological performance of both is established and supported with costs, applications reports and market research information it is possible to make the decision on exploiting the project by seeking capital sanction to move to full-scale production or an intermediary semi-technical scale plant. Alternatively an early return on the research investment may be sought by capitalizing on what has been already achieved in the project, i.e., by undertaking the sale of know-how, patent rights or option rights, etc. These events introduce the exploitation phase.

Phase 4 — Exploitation Phase

Armed with the data secured and summarized in the development phase, the Exploitation Strategy which will have been evolving through the previous phases can be reformulated. Its objective is to explore the optional ways of realizing a return on the research investment which has been made. In this respect it represents the most critical stage of the innovation process after the realization of the initial concept.

The strategy selected depends on the resources possessed by, or accessible to, the innovating company or individual innovator within any inherent financial or technical constraints. In the case of the inventor-entrepreneur, licensing or sale of patent rights to others may be especially attractive, whereas in the corporate organization or company the manufacture and marketing of the product is likely to be considered as the first option, in which the use of existing in-house production capacity may serve to meet the project's needs.

Obviously a number of options exist between the two extremes cited, which entail capital investment and plant design. These are discussed in the subsequent résumé of events which comprise the project as a whole, but the 'conclusion of the innovative process' is marked by the implementation of the exploitation strategy in the form of a presentation of the project as a capital investment business proposal to the investment and business management within the innovating company, with a view to eventual production. Alternatively, where an external company is the potential client for the project, a similar presentation will be made in association with a negotiation of option rights terms for its further development and subsequent manufacture. The latter is of particular priority for the lone inventor-entrepreneur.

An ongoing innovative response to the operation and marketing of product from the plant producing the development prototypes is an essential requirement in most new venture projects and this may entail the negotiation of consultancy or development contracts between the innovators and the client.

The integration of all four phases establishes the framework of the innovation project model in its broadest sense, but some further amplification is now merited to clarify the function of the key events within the various phases. Chapter 2 purports to do this.

2 Key Events of Product Innovation

2.1 EVENTS AS INNOVATION UNIT OPERATIONS

Chapter 1 established that the innovation process structure rests on several clearly definable phases which group together a consistent pattern of related events. This grouping of the key events, derives from the natural subdivision of tasks which are encountered in any attempt to 'take an idea from conception to product realization and eventual exploitation'. The events in question are in effect, the 'unit operations' of the innovation process.

Each event can accordingly be identified with a distinctive operation for which clear objectives can be specified. These objectives define the boundary of the operation or event, but as indicated earlier, the boundaries of one event may overlap those of adjacent events in the same way that one phase overlaps another. These areas of overlap produce a highly interactive situation in which innovative ideas are likely to emerge.

Expanding the Innovation Project Model of Fig. 3, to illustrate the relationship between the various events within each phase, provides us with a 'map' of 'what has to be done' in order to further an idea or other source of initiation, with the overall objective of creating a new business venture.

In identifying the key events, from a range of different 'new materials' type projects, I have found that the recurring pattern seen with the phases, as defined in Chapter 1, is also evident in the case of the events. Also in the context in which we find the events taking place, they can be considered as 'generic' with the phase in which they occur, although differing in detailed content from project to project.

To summarize; the sequence of events is found to fit the analysis of a variety of projects whose structure is based on the broad classification of the four phases, which make up the innovation process.

A typical inventory of events for a new venture project is as follows:

Feasibility events:
1. Idea generation.
2. Experiment.
3. Invention of basic product.
4. Evaluation of product.
5. Basic product definition.
6. Patenting.
7. Preliminary product cost estimate.
8. Feasibility — Project review: plus applications strategy formulation.

Applications events:
9. Base product consolidation.
10. Applications experiments.
11. Applications ideas.
12. Market research.
13. Applications inventions.
14. Evaluation of application products and effects.
15. Application products definition/specification.
16. Patenting of applications.
17. Notional cost estimate of application products.
18. Preparation of development proposal.
19. Applications: Project review and development strategy formulation.

Development and design events:
20. Prototype design.
21. Design of laboratory scale production unit.
22. Applications research.
23. Market research.
24. Notional costing: small-scale semi-technical plant.
25. Notional costing: Commercial-scale production plant.
26. Prototype evaluation.
27. Product cost.
28. Project data summary.
29. Development — Project review, products and process.

Exploitation events:
30. Exploitation strategy.
31. Data selection.
32. Patenting review.
33. Option rights.
34. Consultancy and contracts.
35. Exploitation proposal.
36. Presentation and negotiation.
37. The final goal — project adoption.

Interactivity between events is indicated in the phase/event models, Figs. 4, 5, 6 and 7, by the overlapping of the graphical area of one event with the area representing a neighbouring event.

The feedback of information from one event to another is shown in the models by a closed return loop in the flow-chart construction.

Events are now described in terms of general philosophy in relation to the phase in which they arise. Amplification will be found in the examples and illustrations which are distributed throughout the subsequent chapters which relate each innovation phase to specific projects by reference to case histories.

2.2 FEASIBILITY PHASE EVENTS

Reference to the event inventory of Section 2.1 shows that Events 1 to 8 constitute the feasibility phase as depicted in Fig. 4.

Event 1. Idea

Study of Fig. 4, shows that idea, experiment and invention may continuously interact and indicates that any one of these events can initiate the role of the other two. The permutations of interactions between idea, experiment and invention lie at the heart of the innovation of new products and are accorded special attention in Chapter 4, but the important point to note at this stage is that innovation can start with an idea, an experiment, or an invention, depending on the circumstances which confront the innovator.

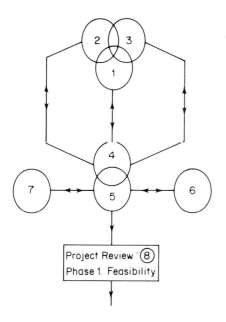

Key
1. Idea
2. Experiment
3. Invention
4. Evaluation
5. Definition and summary
6. Patents
7. Costs
8. Reviews

Figure 4 Innovation project model. Phase 1: **Feasibility**

In the case where experiment or invention provides the initiating step which ultimately yields a new product, they are still dependent on being interpreted through an idea, which postulates the product in relation to applications and this forms the basis on which the innovative chain will be seen to develop.

Ideas are also often instrumental in initiating experiment in a specific field and for pointing out an area of uncertainty in which some inventive solution may be introduced. They are therefore always present in some form at the early stage of the innovation chain, typically postulating what might be attainable if the technological means are available. In other circumstances they postulate the means of implementing desired properties or effects and if these ideas are novel they spill over into invention. As such they represent a most valuable commodity to any innovating company and the prime asset of the lone inventor-entrepreneur.

In the case where ideas are the primary initiator of the innovative process, it is important to appreciate that they can emerge from a variety of sources. They may emerge, for example, from a logical search for a solution to a defined problem. Alternatively they may arise in a highly unpredictable way, which is usually typified by horizontal or transverse thinking and is strongly related to the environmental and experimental influence on individuals, who surprisingly 'see' these factors in juxtaposition with the problem in question and in that process 'recognize' the possibility of a solution in terms of a 'product', providing specific conditions can be fulfilled.

It is especially those ideas in the latter category, i.e., those that possess the element of surprise, which are particularly important to us when subsequently defining any invention, but it has to be appreciated that ideas cannot be patented and hence the importance of invention in the exploitation strategy, yet it is the idea which is the key to the generation of any innovation chain.

Event 2. Experiment

Experiment in the industrial research laboratory is perhaps a foregone conclusion, but the stage at which experimentation is introduced is often dominated by the business interests that the research laboratory concerned are charged with supporting.

In some cases the personnel in a New Venture Research Group will have research targets set for them, which are designed to focus their activity on work which may result in the introduction of new products which will help to expand the 'existing' business. Conversely, broad objectives may be set that seek to promote experimental research into specific selected 'growth' business areas, e.g., fire protection, new materials, speciality effects, etc. Here the researcher has wider scope for exploratory activity which may lead to the identification of clear targets and yield ideas or inventions significantly different from those envisaged when the objectives were first set.

Experimental demonstration of a product or effect at the laboratory stage does not necessarily mean that it will be immediately suitable for economic production by a full-scale process, but the inventor at this point will be alert to future scale-up requirements and whenever possible he will design his experimental process with one eye on its potential for eventual scale-up to commercial production capability. In other words he will attempt to achieve his product using, by preference, well established manufacturing techniques, or at least employ methods of sample production that model the established processes which he realizes might be used ultimately for full-scale production.

Experiment for the lone inventor is inevitably restricted to some extent by resources, but improvisation through necessity is, as we have discussed earlier, not without its own rewards. As the well-worn phrase tells us, 'necessity is the mother of invention'. The result is that the inventor-entrepreneur is almost forced into process innovation, the results of which can enhance the novelty of the ultimate patent property which he can achieve within the life of the project.

Experiment provides ideas for new products, but the converse is also true, that an idea can define an experiment, out of which we then form a secondary idea for a new product.

Event 3. Invention

Invention gives credibility to an idea, or can subsequently give rise to ideas that are expressions of the functionality and application of the invention.

Inventions may be made and claimed as such, without actually demonstrating their practicality; i.e., invention can take the form of postulating what needs to be done to implement an idea.

While ideas can be the prime mover of a new venture, they do not always lead to invention, but it is those new ventures that are based on invention, that are most likely to provide the project with an enhanced commercial value through the protection it offers any potential developer. This value enhancement is the basis for the lone inventor-entrepreneur's business strategy and if the project reaches a successful exploitation stage it serves to ensure that he recovers his research investment with an acceptable profit margin. It also provides the developer with a potential 'Market lead' position on which to secure his return on product development, plant design, construction and marketing costs, etc.

The mechanism for exploiting the added value of invention is afforded by patenting. In summary; it can provide the manufacturer with a monopoly period during which time he can earn a profit from his endeavours while faced with the high overheads involved in launching a new product.

The interrelationship between invention and patenting is of crucial importance to the inventor-entrepreneur since a patented product and/or process is often his chief marketable property on which he must secure his business, typically through the sale or licensing of patent rights etc. It is also important to the corporate research team operating within a manufacturing company, which must take the decision to invest capital in a project. The risk of investing in a project based on a patented product is generally much more acceptable than with a project that lacks such security over its activities, as any approach to a potential client with an invention will quickly reveal.

Further consideration of patent protection is given in Chapter 7, as well as discussion of the role of 'confidentiality' and retention of 'know-how' as alternative ways of securing the commercial value of a project.

Ideas which are incapable of realization in terms of existing technology are the field in which invention can flourish. The inventive phase may itself produce new ideas and so the idea/invention pattern is often repeated to redefine the initial idea or invention and therefore product.

In the limit, the invention can still be claimed to be feasible if the performance of the product or process postulated by the idea is only attainable by means of a laboratory process. A secondary process invention may be needed to achieve a product that is suitable for economic production requirements, without which it is unlikely to achieve a new business venture status.

We can already see in these first three events that the interaction between them is every bit as dynamic as was suggested earlier, but it is evident that the creative idea dominates the initiation stage of the innovation process. This interaction between the three fundamental events of idea, experiment and invention are considered in much greater detail in Chapter 4.

Event 4. Evaluation

The experimental activity and in particular the feasibility demonstration element of it, lays the foundation for the evaluation of the product.

The objective of the evaluation event is to identify and measure all the pertinent parameters that characterize the product, effect or process. In simple terms it asks the question; what have I got? or what has been achieved? From such questioning the innovator compiles information in the form of a product data sheet.

Data sheets, based on a 'new materials' project will be framed typically around the following parameters:

Product description.
Method of sample preparation.
Materials description.
Laboratory tests employed.
Test results.
Application potential forecast, etc.

The list is obviously capable of considerable extension and will vary from product to product. In particular it will be subsequently extended to include data on secondary products and their applications etc.

The evaluation represents a statistical summary of the product's properties and capability for achieving any specific effects and is supplemented by the associated basic processing data. It often includes a comparison with any relevant specifications or standards for products in the same generic group.

The 'evaluation' data sheet serves at this stage to keep everyone working on the new venture informed of the project's status. Subsequent data sheets for market assessment purposes will be more selective in terms of the information to be divulged, i.e., they will concentrate on effects and properties but exclude the process information that shows how the product is achieved.

Evaluation of the invention often leads to a redefinition of what has been invented and the urgent need to revise any patent action which may have been taken prematurely. These aspects are dealt with more fully in Chapter 7.

Event 5. Basic Product — Definition

Once the data sheet has been compiled and evaluated, it is then possible to classify and define the product by a logical analysis of the properties and effects which are summarized in it. At this stage some surprises may be in store as information is revealed that may indicate that the product possesses properties or effects that were not anticipated earlier and which re-characterize the product definition. Since the characteristics were not anticipated, in some cases, we find that they are likely to merit consideration as invention, for invention has the prerequisite that the product's features should surprise us.

In particular, the review of the data sheet may complement our thoughts on 'what has been invented' and assists us in the further identification of inventive elements not previously appreciated. This may hopefully relate the new properties or effects to a field of potential use, in which the asset of novelty enhances its chances of business success.

The innovative function is paramount at this stage since it is necessary to sharpen the idea of a general field of use for the product, into a specific 'use' or 'application'. In consequence it then becomes possible to identify more precisely the range of standards or specifications that must be complied with, if the product is to gain acceptance in the market area, where it is to be targeted.

The evaluation is now reverted to and extended, to determine if the product shows any potential for meeting the market's criteria, some aspects of which may already be satisfied by the information on the data sheet. Other criteria will require further evaluation work in the form of tests and analysis etc, involving collaboration with potential users.

Significantly this further work may again be the source of secondary invention and may displace the initial idea or basic invention in terms of business in terms of business value or perhaps technical merit. Illustration of this will be found in Chapter 3.

The foregoing comments indicate that identifying 'what has been invented' is a dynamic process itself and forms perhaps the most crucial aspect of the whole innovative process where new manufactured products are concerned.

Event 6. Patenting

Before embarking on any market survey or any search for collaborators with whom to evaluate the product for specific applications it is vital to protect the confidentiality of the product and its development in any business arrangement, as has been indicated in the preceding events.

The importance of patenting has been shown in the discussion of invention — Event 3 and the first step in this direction is to take out a patent application with the appropriate Patent Office. The various International Patent Offices provide ample literature for the guidance of the inventor who must ensure that he allocates positive consideration to a patenting strategy at all stages of the project.

The assistance of a Patent Agent is of great benefit to the inventor and needs to be considered at an early stage, as the agent can often make significant contributions to the identification of what has been invented and interpret this in the patent application, so affording the maximum range of options for its ultimate exploitation.

Patenting, of course, provides important protection for the developer of a new product by inhibiting competitors from copying his product or process for up to twenty years, in the case of EEC legislation. This ensures that the innovator has an exclusive marketing position in which any advantage offered by his invention can help him secure a return in his investment on its research.

Event 7. Preliminary Product Cost Estimates

In order to estimate the cost of a new product, and hence determine a preliminary notional price structure, it is essential to obtain some idea of the order of the anticipated scale of production and plant utilization. Such costing presupposes that some information of market size can be approximated and that the product and its manufacturing process is sufficiently clearly defined.

All of this preliminary data is highly speculative, but attempts must be made to arrive at reasonably reliable estimates, since the costing which results will determine the potential competitiveness of the new product in the market, which will have been identified by the product characteristics, i.e. it will provide an indication of 'the ball-park we are in'.

A word of caution is needed at this point. If the notional costing is over-estimated it can lead to the project being thrown away as uneconomical, on what could so easily be a false premise. The point to be made again and again, is that the initial view of a product and the basis for both its costing and application can so often change beyond recognition in the course of events leading to its eventual marketable product definition.

This latter point emphasizes the main difference between the invention/experimental led innovation and that of innovation made in response to an initial market need.

Event 8. Feasibility Phase Project Review and Application Strategy Formulation

Only when the idea and invention have been characterized and shown to be a feasible realization of a new concept, product or process can an evaluation of the significance of any invention be made. The availability of the preliminary data sheet for laboratory samples is of great value in reaching an objective evaluation of the project at this stage, but it must not be forgotten that not every innovative project incorporates invention, whereas demonstrating the feasibility of manufacturing, or otherwise processing a new product, has in all cases to be realized, in order to show that the project can form the basis for a new business venture.

Confidence has to be built up in the project once feasibility is acknowledged to have been attained, by analysing the significance of what has been achieved in terms of parameters such as: how powerful is the effect demonstrated? Or what is the distinctive feature of the product compared with existing products in the field of potential application? Does the product exhibit some novel combination of properties which suggest applications currently inadequately catered for by other products? Does the novelty indicate that the product and/or process of manufacture is inventive and therefore patentable? Perhaps most important of all is: does the innovation indicate to us that a field of business exploitation is open to us in which a diversity of uses may be found with the promise of several secondary product outlets with added value potential?

Data sheets afford rapid comparison with potential or established competitive products and enable a quantitative assessment of the product to be made. Typically a table of material strength properties can be rapidly compared, to establish the scale of any advantages or disadvantages that the product may have relative to other materials, in which strength also characterizes its usefulness.

Clearly a knowledge of a wide range of markets and an awareness of applications is needed by the innovator or inventor if the value of any invention or opportunity for generating a new business venture is to be recognized at this stage. This is where the collective knowledge and wisdom of other members of the project team becomes invaluable in discussion of properties, effects, design features, etc., all of which aids the evaluation which determines the level of future resources to be allocated to the project.

At this point it is timely to seek the view of the marketing organization who can initiate preliminary market surveys on specific areas, some of which the inventor and project team may have identified. This will be further amplified by the marketeers themselves who have a range of experience of market opportunities which differ in emphasis and content from those identified by the research team, but which can often become a starting point for further product innovation.

If the review is positive, it will select where to target the application of the product. In making this choice, the market area where an application can have most prospect of meeting the technical requirements, has to be balanced with that which offers the largest market opportunity, but the deciding factor may rest on which field of application will lead to the most cost-effective product.

It must be said that targeting of applications are usually implemented by identifying those applications which are already catered for in the market place by existing and competitive products. The innovator's response is to seek to identify new applications for which the new product offers unique benefits in cost or technological terms. Examples of this generation of new applications products appear in the relevant chapters.

In mentioning cost-effectiveness as a selection criterion at this stage it is important to recognize that many projects are launched with products that are loss-leaders, i.e., they are certainly not cost-effective products, but by marketing them at an uneconomic price they are used to test the market and open up other areas of application in which the cost-effective criterion can be applied successfully.

Clearly the identification and selection of applications is a vital issue which must be initiated at this review stage, unless applications have already emerged at an earlier stage i.e., in the feasibility phase. In either case an applications strategy must be formulated in which the constant feedback of market data and technological advance with the product can be assimilated in support of the search for cost-effective applications.

2.3 APPLICATIONS PHASE EVENTS

Events 9 to 19 constitute the applications phase as shown in Fig. 5.

Event 9. Consolidation of Basic Product

Once the decision to advance with the applications phase has been made; technological improvement, costing, patent application and market research will all be given a fresh impetus in the effort to secure a business potential for the base product and to reaffirm or consolidate its properties and effects. In addition, improved laboratory production of samples will be needed to furnish the market research and the applications experimental work.

During this process of consolidation and the allied applications research into the utilization of the base product, it will soon become evident that the product has certain limitations which restrict its applications potential. As a result, attention will be focused on overcoming any shortcomings identified and this in turn leads to the innovation of variations of the base product specification often culminating in the introduction of a product range. Success in this sphere of adaptability is the aim of product consolidation, i.e., it seeks to ensure that the base product will be adaptable to as wide a range of applications as possible while endeavouring to keep the variations in its specifications to a minimum.

Event 10. Applications Experiments

We come now to the role of experimentation as the hub of applications research in which secondary innovation is frequently the outcome. Experi-

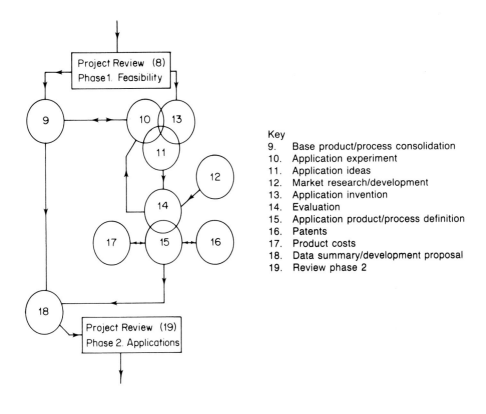

Figure 5 Innovation project model. Phase 2: **Applications**

mentation plays an important part in enabling the base product to meet a variety of applications as described under Event 9, but where it cannot meet the requirements of an application, the innovative process extends into the generation of secondary products.

In some circumstances, effort is made to utilize the base product as a starting material or component of the secondary products with a view to increasing the overall production requirement for the base product. This will lead to improved economy of production costs through process scale-up.

The applications research, as we shall see, is closely aligned with the market research activity but is motivated by the innovator who may be the inventor behind the initial idea and basic invention. Alternatively it may be the province of an innovator specializing in secondary invention, i.e. in the exploitation of a basic invention which has been introduced by others. Whichever is the case the applications research starts with the generation of ideas for uses of the base product in selected fields of application where its properties can be economically and technically exploited, if required, by secondary or applications products.

Fundamental to any applications research is the need to introduce quality control on all product samples used in the programme. It is at this stage that the innovator or inventor reaches out to potential collaborators to explore and identify the application areas for the product in which 'user' participation can be gained, as a result of which specifications for the product can be tailored to requirements.

Testing is a key area of collaboration in which the potential user's contribution is essential, in many cases enabling the innovator to ascertain the key properties which will be required by the product in a specific application. This represents a vital step in ensuring that the base product or secondary products will be properly prepared to meet the wider challenge of the market and of making an entry into existing processes, products, systems and business in general. Testing is discussed in more detail in relation to the prototype product, under Event 22 in the development phase later in this chapter.

The probability of introducing secondary invention at this stage is obvious and as pointed out earlier, it will be evident that such inventions will need to follow a similar innovation process route to that used for the base product.

From this constant requirement to interact closely with the potential market it is evident that the applications specialist must keep in close contact with his marketing colleagues at this stage since, should the initial technical collaboration have been secured with a potential customer for the new product, it will be necessary for the marketeer to ensure that the overall business interests of the project are recognized in any formal arrangements between the innovating party and the collaborating party. A better arrangement is gained if the 'initial' approach for technological collaboration on product evaluation in the market is made in partnership with the marketeers. Do I hear my old colleagues of the marketing team raising a faint cheer?

The introduction of the marketing and business strategy factor may, unfortunately on occasions, inhibit some collaboration with companies who can greatly assist in the technological or scientific advancement of the project, but who regrettably turn out to have vested interests in the project taking a course which serves their own business interests to the detriment of the new venture. It is a difficult area which needs effective partnership between the scientific or technological members of the research team and the marketeers in order to ensure an effective communication with the collaborator.

Chapter 5, gives more extensive consideration to the applications/marketing/business strategy interface.

Event 11. Ideas for Applications

The generation of new ideas aimed at extending the usefulness of the base product depends on the ability of the innovator to recognize the distinctive features of the base product in terms of physical properties, effects, processability in new 'end forms', cost effectiveness, etc. This knowledge has then to be used to obtain a technocommercial fit into the market place with the product adapted to meet the specific requirements of respective applications.

Occasions arise when the ideas for applications cannot be implemented with the base product even in modified form or specification. The result of such ideas is that they can then lead to the invention of secondary products and processes which no longer depend on the 'parent' project, as intimated earlier. These secondary innovative steps constitute a new project, which makes a 'shift' in the innovative chain by redirecting the energies of the research team or individual researcher in pursuit of new targets, providing the 'shift' can be shown to offer the prospect of establishing a successful business opportunity.

The introduction of such radical changes as a new project may involve the abandonment of the initial project. Alternatively, the new project may run in parallel with it, providing both projects fit the business strategy and can be implemented with the resources available to the new venture enterprises. In some cases these secondary projects have such affinity with the initial project that they serve to complement each other in the research and/or marketing fields.

Ideas are evidently the most powerful and fundamental factor in the innovative process.

Event 12. Market Research

The preliminary market survey can only commence in earnest when the data sheet for laboratory samples has reached the stage where sufficient information of appropriate reliability is available, to form the basis, for opening a dialogue with potential customers or collaborators.

The survey will play a key role in extending the interest in the product to those who can contribute to its usefulness by collaborating on the definition of product targets and specifications to suit their own applications.

At this stage the marketeer will begin to undertake an assessment of the overall techno-commercial structure of the potential business that may arise from the project. This will become of increasing importance in determining which potential collaborators should be adhered to and which customers might provide the best outlet for the product at the differing stages of eventual market development. It has to be noted that collaborator and customer may be the one and same party.

Event 13. Application Inventions

Whenever an idea is generated for utilizing the basic product in some application, it usually carries with it an appreciation of the extent to which the product is able to match the techno-commercial requirements demanded of it. Where the fit is unsatisfactory it immediately points to a problem for which an innovative solution may be essential. It is in these areas that application inventions arise and secondary invention relating to the initial base product may also lead to a revision of the latter. Precisely how this secondary invention occurs is difficult to generalize on but the various project examples provided in succeeding chapters reveal to a marked degree, exactly how it does occur in specific and detailed instances.

Event 14. Evaluation of Applications Products and Effects

One of the features of applications research already touched upon, is that it must directly involve potential customers at an early stage, in order to resolve questions of specification and quality control. Without recourse to such collaborators the product cannot always be adequately assessed or evaluated. Such specifications form the basis for further development of the product or precipitate the secondary inventions which enable it to meet the requirements of the application concerned, on the assumption that the initial applications product is inadequate.

The importance of this stage to the project's success cannot be overstated, since it is at this time that the product proves itself to the potential user, or fails. In the illustrations drawn from the various projects, the struggle to meet specifications is evident, as is the equally difficult task of ensuring that vital criteria laid down for the way in which a product can be employed are adhered to. I recall one or two instances in particular where the failure to use the product as prescribed led to collapse under test conditions, much to our embarrassment.

The sequel to all of this collaborative activity is an applications product data sheet, as distinct from the data sheet for the base product. This presents all the characteristics of the applications product related to the specific use and hopefully provides convincing evidence of its suitability for the application concerned.

Event 15. Definition of Application Products

Specification requirements of application products must include both quality control and test criteria at the manufacturing stage and at the user or application stage. Clearly the 'controls' at manufacture refer to product and process specification, whereas those at the 'user end' refer to the way in which the product is utilized in a specific application. Only by a stringent regard to both criteria can a manufactured product build up market confidence, especially if the performance of a product can be changed by the assembly or method of use. Application product definition must reflect all these aspects.

The implication of the above is that the product data sheet must be extended by reference to the user's specifications and test criteria in order to show that the product will perform in services conditions to the user's requirements.

Reference must be made also to the cost estimates of the application products based on anticipated market penetration levels, since these may demand revision in the light of information gained from the market place and changes necessitated in the product specification and design. It is often found that in association with this recycling of data and specification, new ideas and further invention arises. Application product definition draws together the technological, marketing, cost data and patenting in an applications proposal which forms the basis of the attempt to establish the application markets.

Event 16. Applications Patents

The comments made for Event 6, regarding patenting of the base product apply equally to the application products, but they will be focused now more sharply on the potential use for the product opposite a particular application. This can lead to the introduction of invention and patenting of the way in which a product is used as well as how it is manufactured, i.e., applications patents need to encompass a wide field to support the eventual application and marketing effort.

Event 17. Cost Estimates of Application Products

Costs at this stage are of crucial importance since the potential customer's collaboration throughout the applications phase is essential and it is the competitive element that may determine the extent of continued interest in the product.

For the reasons given, it becomes essential to tighten up on the estimation of costs, to a degree that can be sustained when arriving at a starting price for the product, which can then be used to establish an exploitation base for the project.

Event 18. Development Proposal, Laboratory Scale

Everything that has been achieved so far in the innovation process has been based on laboratory scale evaluation and testing, but it is now imperative to demonstrate that the respective products, i.e., base product and applications products, can be produced by a laboratory process that models in principle the full-scale processes envisaged in all critical respects, i.e. to show that the samples produced will confirm that practical quality control can be exercised in their production. In achieving these targets it must also be demonstrated that the properties and effects attainable in the initial experimental or feasibility phase are retained or 'improved'.

The data needed to make the decision for starting the development phase is: technical, marketing, costs and patents. The development proposal brings this information together and seeks to display that appropriate laboratory processing facilities can be realized to permit the production of sufficient samples, made expressly to satisfy a variety of standards and test criteria. A level of compliance with these is required to give 'indicative' performance values as well as 'properties' and provide a basis for judging what probability exists of any future production version of the product satisfying the same or related standards and/or specifications.

Event 19. Applications Phase Project Review and Development Proposals

This review is concerned with making the decision to proceed to the start of the laboratory development phase of the prototype for the commercial product, in both basic and applications product categories. Two of the most important factors to be considered in more detail are:

(1) Has sufficient verification of the properties and effects been obtained with the laboratory samples, to permit extrapolation to a prototype product design for the product range under consideration?
(2) Has the laboratory experimentation on sample manufacture provided sufficient information and ideas on which to base a small-scale laboratory plant with sufficient capacity to meet the needs of the next group of evaluation tests which involve some degree of scale-up in sample size or complexity?

It will be obvious that embarking on further research investment brings a demand for assessment of the project's chances of realizing a business success. The lone inventor will attempt to avoid any over-sophistication at this stage by innovating in terms of effectively modelling the full-scale process envisaged, only with respect to those parameters that are critical. His alternative course is to make early attempts to convince one of the collaborators that the time has come to acquire patent rights to the manufacture of the product and undertake its development phase, using their own resources and drawing upon the inventor for consultancy support, as required.

Whichever course is followed, the prototypes must eventually be produced to satisfy an agreed quality control standard which is considered to be attainable also by a future large-scale production plant.

2.4 DEVELOPMENT AND DESIGN PHASE EVENTS

Events 20 to 29 constitute the development and design phase as shown in Fig. 6.

Event 20. Prototype Design

The prototype product provides the model for the final full-scale production product. This entails a complete reappraisal of the laboratory sample product in terms of its properties, appearance, functionality and method of production. In the latter case it will usually involve making a change from simple batch type processing or manufacture, to one which more closely resembles the mass production process envisaged for the full-scale production plant.

To illustrate the need for the design of the prototype product; a relatively small laboratory sample of say, a composite material, is generally adequate for ascertaining most of its key properties, from which its usefulness can be broadly assessed, i.e. it may indicate for example that the material could be exploited as a fire barrier wall or ceiling panel. However, because of the small size of the

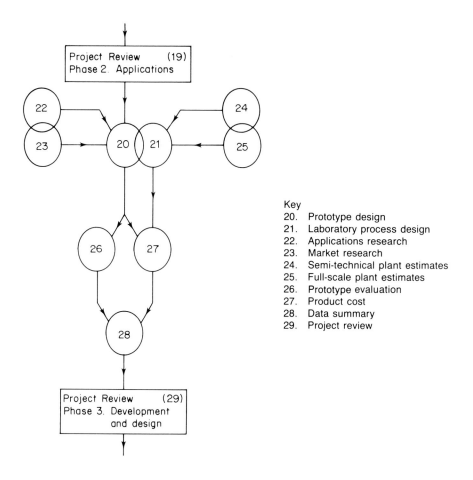

Figure 6 Innovation project model. Phase 3: **Development and design**

samples available it is unlikely that adequate verification of its functionality for the indicated applications can be determined against the wide range of statutory fire tests which must be complied with, if the product is eventually to find acceptance in applications such as arise for example in the building industry. This is because the tests which are stipulated for these applications demand samples which relate closely to full-scale products.

In the example cited, the prototype design will change the emphasis of the innovation process from simply demonstrating a new material to one of manufacturing and end processing the material into a new product form for a specific application, i.e., as a building panel which meets the standards of the building industry with respect to size, properties, appearance and other appropriate specifications.

This scale-up to the prototype and the associated scale-up of test samples represents a significant development effort and the time and resource allocation needed to implement it ought not to be underestimated.

It is evident that for the design to be effective, an intimate knowledge of the potential market and application requirements is essential so that the design can reflect these. The marketeer and market research team provide a key input in collaboration with the applications specialist or inventor at this point. The final outcome of this collaboration is that the prototype product will be designed to meet the specifications which have been identified and be representative of the anticipated full-scale production product in terms of properties and many other respects, in which case it can be used to evaluate performance against respective applications test criteria as representing the full-scale production product.

In many statutory tests the specification of the sample, for submission to the test, may be such that the prototype product sample required may be less complex or smaller in scale than would be demanded of a commercial full scale prototype. *As a consequence of this the innovator can then set his initial prototype specification to meet the test requirement sample details instead of attempting a single leap from laboratory bench, to produce a sample of commercial scale or complexity.*

A typical example is again provided by the fire barrier building panel which must meet certain statutory requirements with regard to 'the extent of surface spread of flame observed when the panel is subjected to a defined ignition source'. This demands that samples of a specific size be produced or cut from larger production material, i.e., requiring much larger samples than the early laboratory samples used in the feasibility phase, but smaller than the envisaged final product, which would be offered to the industry. Such samples often involve changes from the initial specification which characterizes it as a prototype application product, but these changes must be consistent with the future full-scale production specification for the final commercial product.

To summarize; the prototype design stage involves translating the laboratory sample into a product specification that can be implemented by full-scale production processes, but which will in many cases 'initially be produced' using laboratory resources which model the full-scale process. This may entail the use of batch production methods instead of the anticipated continuous methods providing that the end result, in terms of product characteristics, properties and appearance are still achieved and are consistent with the full-scale product target specification.

In addition to satisfying the performance criteria for the applications for the product, the prototype design must also cater for other considerations. Any new product must compete with other products which may already possess both aesthetic and other more functional features in their design. Clearly the prototype design stage ushers in an environment of change, i.e. innovatio through design, may assert itself and lead again to secondary invention. In this respect it is an advantage if the innovator of the new product possesses design experience and it is one of the obvious reasons for the prominence of engineers among innovators of manufactured products.

Event 21. Laboratory Process Unit Design

The possible need to make prototypes on laboratory type production units has been argued in Event 20. This step represents a small but important scale-up of the laboratory production of samples in which methods of exercising quality control plays a dominant part.

In the prototype design stage we indicated that the samples produced primarily for test purposes must be representative of production samples. To satisfy this requirement the quality control introduced at the laboratory stage must also reflect that type of control which could be used in full-scale processing and this may necessitate the design of a new laboratory process facility. The samples produced by such a unit will be employed in market research and market development as well as in the test programme which aims at gaining product certification for specific applications.

The specification for the laboratory process unit is not to be confused with that required for a semi-technical plant which will eventually supersede the Unit and provide the scope for fully evaluating design parameters for the ultimate full-scale plant and meeting a significant increase in the scale of product evaluation and market development from its significantly increased production capacity.

Event 22. Applications Research with Prototypes

Although the main thrust of applications research is a distinctive phase of the innovative process as has been already pointed out and shown by Fig. 3, the application activity overlaps into the development phase activity and the development and application of the prototype is the focus of this interactivity.

This is most evident where part of the evaluation of the prototype involves collaboration with a potential user, an interested research organization, or a testing and certification establishment and calls for close cooperation with the innovator leading the applications research. Among the tasks which require special attention as indicated earlier, is that of ensuring careful quality control analysis of the samples provided and close observation and interpretation of each test, broadly as discussed under the previous events.

It is during the quality control monitoring stage that product 'limitations' become evident and spark off ideas for possible ways of improving the product in order that it can then meet the test criteria. On occasions this leads once again to secondary inventions which will increase the market potential of the product or at least enable it to hold its place in the competition for the targeted application.

Event 23. Market Research with Prototype

Some of the market research will have been carried out with the initial laboratory samples, but the availability of prototype samples, i.e., those representative of full-scale products, presents an opportunity to follow up these first encounters with the market place often in collaboration with the applications specialist, armed with samples that exhibit a more complete picture of the product's capability and other aspects such as aesthetic appearance, professional finish and design features, etc., not evident in the earlier laboratory samples.

The fact that the prototype models the full-scale product provides the potential user with a new perspective of the relevance of the innovation he is faced with and stimulates a reassessment of the value of the products to the company in question, encouraged by the actuality of a prototype which can be seen and handled.

Valuable updating of market potential is the result of this stage of the innovative process.

Event 24. Notional Cost Studies – Semi-technical Plant

The market sampling figures from Event 23 will determine the size of the semi-technical plant and form the basis for calculating capital investment requirements for it and facilitate the calculation of the expected product cost for various levels of plant utilization. Product price may at this stage have to be marked down in order to enter the market in the expectation that initial losses can be recouped subsequently when reaching full-scale plant production.

Event 25. Notional Cost Studies — Full-scale Plant

The data available for considering the size of full-scale plant required to meet expected demand is inevitably more speculative than applies for the semi-technical plant, but the estimate is particularly important because it greatly influences the commercial product price, since product cost is a function of production scale. The estimate, of course, determines how much capital is needed to initiate the new venture's production capacity.

The subsequent building and operation of a semi-technical plant, of course will provide valuable information with which to up-date the full-scale plant and product estimates, but this activity, which lies in the engineering design field is outside the scope of this book. What is still relevant is that operation of the semi-technical plant and to a lesser degree, the full-scale plant, will continue to provide an area where innovation will be required on both product and process in response to the market penetration stage of the product exploitation.

This latter area of activity is usually identified with the technical service support normally set up when the product is first launched on to the market. It represents another important opportunity for the innovator to make his contribution, but the emphasis in such departments is predominantly on guiding the user in the most efficient way of utilizing the product and on responding to problems that may feature in quality control and specification variations.

Innovators may consequently find that their career pattern may lead them into technical service situations as they continue their contribution to the project in which they have played a creative role. This may be appropriate for some, but for other innovators the close proximity to basic research will be found to be the greater stimulus for their aptitudes.

It is worth re-emphasizing that the cost studies of the semi-technical and full-scale plants are not intended to form the basis for capital sanction requests at this stage, but to provide the essential cost data from which a potential price structure for the product can be estimated. On eventually turning to the exploitation phase this product price estimate will be of vital importance in any negotiation.

Event 26. Prototype Evaluation

Prototype evaluation is an integral part of the applications research programme which overlaps into the development phase. It usually includes several categories of tests which are used to progressively support the innovation of the products and process throughout the life of the project. Four major test categories which feature are:

Indicative,
Advisory,
Mandatory,
User.

The first of these, i.e. indicative tests, concerns the initial testing of a product which is usually carried out in-house, i.e., the innovator evaluates 'samples' from the earliest stages of the project, including both laboratory samples and prototypes using tests of his own devising or selection. This provides him with an indicative picture of the product's potential for meeting more rigorous subsequent test requirements in the other three test categories. A considerable proportion of any project team can be taken up in this stage of the work. In the case of the lone entrepreneur he will often need to carry out the tests as part of his own research activity and will often improvise the test procedure to effect the maximum economy on expenditure and time, but this is itself an incentive to achieve simplicity in eventual quality control of the production process.

An important aspect of the indicative tests is that they provide a means of rapidly and cheaply screening numerous versions of the product before selecting the relatively small number of samples for submission to the more critical and expensive tests.

Advisory tests are usually incorporated in international product standards or design codes and serve to form a comparison basis for the researcher while setting a quality control standard for the benefit of the potential user. In some countries compliance with the standards or tests is mandatory for specific products.

Mandatory tests are also introduced by Government institutions in some cases before any manufacturing standards have been established. This represents a way of impelling changes and improvements in products to meet various needs of society. Typical instances are in the USA where automobile exhaust emissions were the subject of legislation with the result that systems for emissions control were invented and developed on a wide scale.

User tests, as the name implies involves the potential user of the new product in setting up his own acceptance criteria and tests for a product which embraces both basic properties and those associated with the application environment for the product, i.e., in the latter of these it entails tests which relate to the product's performance *in situ* with his own process or product of which it may be part. It follows that these tests will usually be carried out with his own resources at his own premises. Where this is the case it is essential to secure access to his test programme and here we find the maximum need for the innovator and marketeer to work closely together to ensure that the prototype product evaluation is conducted in accordance with the quality control specifications that have been set for its deployment and use.

Features of the evaluation activity are that compliance with the test requirements of Statutory Bodies transforms the status of a product and facilitates its exploitation by increasing market confidence. During each test period it will be evident if the innovative advancement of the product has been secured by comparing product specification and performance. It is essentially the exposure of weaknesses as well as strengths in the product's performance that results in still further refinement of it and often leads to secondary inventions in order to meet the test requirements demanded for a particular application.

Much can be learned by the joint deployment of technical innovator and marketeer during the user tests working in close collaboration with the potential user. Examples given elsewhere illustrate that this includes performance characteristics and aspects of commercial strategy, both of which aid the product exploitation.

Clearly it is beneficial to carry out all categories of evaluation, since this will yield the maximum data on the 'properties' of the product and achievable 'effects'.

Event 27. Product Cost Studies

Within a corporate research organization the research economics team has the resources to assess the scale of both future semi-technical and full-scale process or manufacturing plants, using data received from the applications research and market research activities.

What is perhaps one of the most critical events after the initial invention has been made can now be undertaken i.e. the research economist, the inventor-entrepreneur, or others, will prepare cash-flow calculations for the project on the basis of available data and this will reveal the potential price structure of the product or products, as well as capital and revenue costs for the construction and operation of the respective manufacturing plants.

To achieve these estimates, it is essential that the notional engineering design study of the process plants has been carried out in sufficient detail to identify all of the main plant operations and to facilitate the drawing up of a notional specification for them. Such design may constitute part of the innovation process at this stage and the inventor may of necessity be forced to innovate in terms of a process design himself in order to produce his product or effect and interpret it in cost terms.

In some cases the design study will require the major resources of a design organization, who produce notional process plant designs, as the preliminary to the submission of cost estimates for sanction of capital investment programmes by the investment managers.

In relation to the above and Event 22, it can be seen that the notional design for a semi-technical process plant and a full-scale commercial plant must be considered separately since each have different functions to perform. Events 26 and 27 refer.

Event 28. Project Data Summary

The evaluation of the prototype and the summarizing of the product and processing cost data, with the addition of the latest estimates of market potential, on which the production scale estimating is based, is finally brought together in a project review document.

Integral with this document we find the data sheet which lists all the key properties of the product in its prototype form and the results of any application and evaluation tests of both base and secondary products. This comprehensive summarizing of all that is relevant to the market potential of the product and therefore of the project as a whole, provides the substance of the development and design phase review which seeks to prepare the project for its next and final innovation stage, i.e., the exploitation phase.

Event 29. Development Phase Project Review

As Event 28 intimated, this review contains the critical decision point for the innovation chain which asks: Is the product sufficiently developed and supported by the relevant cost, test and scientific data, etc. to merit the start of the exploitation phase, with a view to securing adoption of the project by others who would either scale up the development with the intention of production and sales or undertake some alternative initiative in the product's exploitation?

The review raises many question which are dealt with in detail in subsequent chapters, among the most important being: does the prototype product or products meet the cost/effectiveness criteria demanded for entry into those areas of the market for which applications have been identified and evaluation successfully carried out?

Assuming that the cost is competitive, that the product has been shown to possess novelty and the process has demonstrated that it is economically practicable, the prospects for turning the project into a new business venture could be high, providing the cash-flow studies also show that the venture will be expected to reach the break-even point within the resource time limit and yield a return on investment as profit at an increasing rate, subsequently meeting the profit estimate levels originally assigned.

Within the corporate organization, once a positive support for advancing the project further is established, the route must be charted in the exploitation phase. The lone inventor-entrepreneur has similar criteria to satisfy in order to establish a strong negotiating position in seeking a client who will undertake the final development, design, construction and marketing stages to take the product into a sales portfolio.

2.5 EXPLOITATION PHASE EVENTS

The events of the exploitation phase, shown in Fig. 7, provide the completion of the innovation project model.

Event 30. Exploitation Strategy

Reformulation of earlier and preliminary exploitation strategies in the light of the development phase activity is of vital importance for successful exploitation of the project since the experience gained from the development and evaluation of the prototype products will have progressively extended the range of contacts and collaborators, among whom may be the potential client with the interest and capability to exploit the commercial value of the project.

Both market research and applications research, conducted throughout the application and development phases, will have made their input to Events 28 and 29, i.e., the project summary and review of the development phase. This now represents a major factor in the formulation of the exploitation strategy when allied to the update of the patent application position. Chapter 7, discusses the optional forms of exploitation strategy, through which the drive to realize the adoption of the project as a business venture is achieved.

Event 31. Data Selection

Once the overall exploitation strategy has been reappraised in the light of the latest information, the next target is to prepare for an exploitation proposal to potential clients.

The first requirement is to make a selection from all the scientific, technological and commercial data which has been generated by the project, against the criteria that it should adequately convey the current status of the project

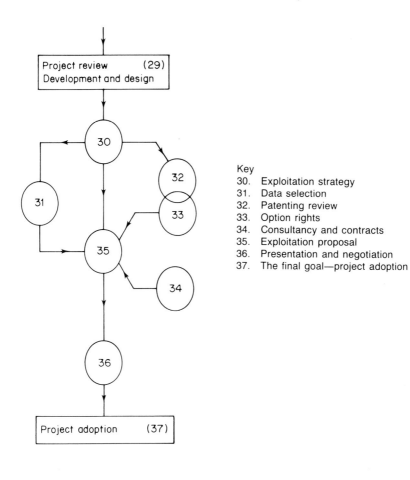

Figure 7 Innovation project model. Phase 4: **Exploitation**

while observing any constraints imposed by the chosen exploitation strategy.

Clearly this will draw upon the product data sheet and the market review and application test data which will have already influenced the exploitation strategy formulation. It must be emphasized that the data selection, which will have been summarized at the end of the development phase, i.e. Events 28 and 29, has to pay particular regard to excluding selected details of the product and process in order to maintain a strong confidentiality position during the exploitation/negotiation phase.

Event 32. Patent Review

Complementary to the data selection we find the need for a review of the current patent position on the project. This entails questions of the present status of the patent application and the strategy adopted regarding which aspects of the project will be kept 'confidential' at this stage in the form of retained 'know-how'.

Particular regard must be given to base product, base process, applications products and application processes.

Event 33. Option Rights

In this event, which is closely associated with the overall patent strategy, option rights form an important part of the negotiation spectrum. Chapter 7, provides a fuller explanation of their role in negotiation, but suffice to say here, that they represent an important means of securing a financial return on the research for an innovative project and are of prime importance to the inventor-entrepreneur when seeking to exploit his work.

Event 34. Consultancy and Contracts

Before we can complete the preparation of the exploitation proposal we need to give thought to the potential that may arise in any negotiation for establishing a consultancy role with the potential client and/or setting up a formal research or supply contract to assist his evaluation and hoped for exploitation of the project in terms of further development and the design, construction and operation of a production facility or plant.

Linking any contracts or consultancy with the negotiation of option rights is highly desirable in order to avoid misunderstanding over questions of collaboration in future activities where assistance is needed by the client. Examples of the latter appear in later chapters. It must be emphasized that while linking these in strategy it is essential to negotiate quite discrete terms for all three arrangements.

Event 35. Exploitation Proposal

The Proposal is the means of communicating to the potential client what is on offer. While it depends primarily on a written and formal presentation document which sets out all the preselected information already indicated, it must also include samples of the product and in particular those versions of it that relate specifically to the areas of known interest of the 'client'.

Of prime importance: the document will set out the range of options open to the 'client' for his inspection and form a basis for discussion.

The question of option rights fees has to be carefully weighed before any negotiation takes place, but is better left out of any formal document at the first round of negotiation, as the level of interest and value of the project to the client must be first ascertained during the initial meeting, at the negotiation stage.

Event 36. Presentation and Negotiation

The end is in sight! All is prepared for the effort to convince the client of the project's potential for forming a promising new business venture with which to expand his existing business or create a new one.

The presentation involves documentation, backed up with visual aids and culminating in the display of a prototype product whenever possible. In fact it is the visual contact with the new product that makes most impact. The visual aids, support this effect and the documentation is often treated as a future reference paper with which to check out particular details of interest after the meeting.

With the foregoing in mind it is obvious that the presentation environment and facility for product and visual aid display is of great importance. Much of this requires careful planning and a clear and logical timetable of events on which to make the presentation. Chapter 7 expands on this more fully.

Event 37. Project Adoption

The final goal is the adoption of the project as a result of the negotiation on the basis of terms which are agreed, often subsequently to the initial negotiation and usually after several sessions have elapsed. Obviously it is imperative to retain full ownership of any patent property until terms are finalized and this requires that some formal Confidentiality Agreement is entered into by both parties before the negotiation stage is entered upon.

Providing a full disclosure of all the pertinent facts needed to enable the client to assess the projects potential is made, a harmonious collaboration can be secured and hopefully the project will be adopted by the client on terms, such as those which are detailed in Chapter 7.

Application examples of the innovation model are provided in all of the succeeding chapters with special reference to amplifying the activity of the various phases of the innovation process.

3 Innovation and the Innovator

3.1 THE INNOVATOR'S ENVIRONMENT

In Chapters 1 and 2, the concept of the Innovation Project Model, as expressed by Figs. 3 to 7, has been established and from this we saw that the innovation of manufactured products, such as 'new materials', involves four major phases of activity i.e., feasibility, applications, development/design, and exploitation. The model has also shown that the dynamic coupling of various key events make up the respective phases of the innovation chain.

When such coupling of events occur, depending on their specific role in the 'process', they take the form of either a lateral or transverse link in the chain, as Figs. 4 to 7, indicate. The transverse coupling involves adjacent events at a common node point on the chain, whereas the lateral coupling represents the transmission or recycling of information between events along the chain.

Each of the new venture projects that I have examined fit the broad structure of the Innovation Project Model, but they differ in the particular sequence and combination of events that occur and it is this that characterizes the innovative chain which fits a particular project. An example of this is provided in Chapter 4, where the possible options relating to the sequence and combination of events which govern the earliest stages of the innovative process are examined in detail, with reference to Fig. 10.

The motivation for the events to couple, exchange information, or interact upon each other is influenced by several factors. Among these is the *degree of awareness* of the innovator with respect to events that have already transpired, or which are simultaneously occurring elsewhere on the innovation chain. In the limit, innovator awareness, includes the anticipation of future events and their potential contribution to the current focus of attention in the chain.

The extent of this *awareness* in many cases is due to the innovator following a strictly logical line of enquiry and development, but may also arise subconsciously in the form of a *creative response* to the information exchange taking place between events, involving both lateral and/or transverse interactions between them. In my experience, these creative responses may be triggered by internal situations, or by completely external influences.

This points to the importance of managing and organizing a project in a way which is conducive to the establishment of an internal environment which is sympathetic to the way in which the innovator works within the research team, in particular by providing a logical progression and review of the innovation process phases and by exposing the innovators to the stimulus of the cross current of experimental enquiry and debate within a range of projects, which of course includes the particular new venture which is the focus of interest.

Much of the effort expended by the new product innovator extends to collaboration with potential customers and various collaborating bodies etc. These represent the external environmental influences on the innovator and are the result of pursuing a specific exploitation strategy and associated product applications which are determined by the type of business in which the product will be expected to find its market. The latter may introduce the innovator to areas of technology with which he may have previously been unfamiliar and provide information that acts transversely with the ideas generated internally, producing new or secondary ideas for new products.

'Getting the environment right', can be seen to depend on a management strategy that progresses a project in a logical way and exposes the innovator to both internal and external environments by ensuring a high degree of internal information exchange supplemented by maximum exposure of the innovator to the market environment, especially through application and market development stages.

The foregoing remarks indicate that the innovation process is quite complex and suggest that it will not be comprehended in the same way by everyone, including those intimately engaged in the process itself. This in turn can give rise to differences of understanding about the respective role of researchers, marketeers, etc.

An example of how incomprehensible some people find the innovative function to be in a new venture project, was brought home to me when, during a project review, one of the marketing team commented that the inventor concerned seemed to have had the 'luck' to find that a particular process operation could result in the invention of a new product. To my mind this showed that he failed to realize that it is not simply luck, *but the result of a developed sensitivity to the juxtaposition and significance of many events, often related to experience in other fields that provides the creative idea for the inventor*. Nor did he appear to recognize that a disciplined logical advancement of the creative idea is implicit in ensuring its eventual practical realization.

That some are creative in their thinking may be an irritation to those who are not, we often say 'why didn't I think of that', but the fact that some react to their world in a way which is 'creative' should not be construed as a reflection on the capabilities of those who do not. Both have roles to play in achieving project success and each possesses different attributes.

As in all aspects of management, the need to secure good liaison between participants of a team is of paramount importance for as Parker (1980) states, in ref. 10, Section 5, *Organising research, development and design for innovation — Guideline 5*; that 'Marketing, innovative groups, and manufacturing groups, rarely share the same perception of their responsibilities. Some conflict is inevitable, but it must be curbed. A little stress helps creativity, too much destroys it'. Clearly defining who is in a new venture team is a crucial decision for management.

We have attributed considerable importance to the part that management plays in creating the environment in which innovation may be pursued, but another key factor is the professional environment created by the influence of the innovator's own 'discipline'.

The observation I wish to make is that the innovator's discipline, conditions his response to problems or design briefs; e.g., the engineer working in a mechanical engineering stress analysis design capacity in a design group is *stress* orientated in his innovative responses, whereas the engineer engaged in an aerodynamicist's capacity in a research laboratory is likely to respond to a problem in terms of *experimental aerodynamics*. In the wider scientific field a chemist specializing on research with inorganic materials will be conditioned to some degree to reacting to phenomena with ideas based on the structure, properties and effects attainable with *inorganic materials*.

Clearly the environment of thought in which an innovator specializes is the one in which he is most likely to 'recognize' intuitively that some element of it provides a *conceptual fit* with information which he receives from other observations, references or stated needs. It is in this process that I think innovation and at times invention occurs, often as a result of synergy.

A particular example of the different ways in which an *innovative response* is conditioned by environmental influences occurred to me early in my career while working as a stress analyst in a design department. Part of the work entailed checking the mechanical design of various types of chemical plant structures, including tall towers such as gaseous effluent stacks or chimneys and distillation columns. In this instance several tall distillation columns were to be stressed-out to determine their safety levels under wind loading conditions.

The critical factor in the design of such towers, is that the working stress levels due to the loads imposed by sustained high velocity winds on exposed sites, must comply with the safe stress levels demanded by the design specification for the selected materials of construction used.

My approach to the problem as a stress analyst at the time was to consider the nature of the forces on the structure and then design the proportions of it to achieve inertial resistance to the deflections which would keep the stress levels within acceptable or specified design limits.

The stresses which can arise on column shells of this type are due to the significant deflections which can occur under resonance conditions from the influence of sustained steady state wind conditions. This arises as a result of the well-known phenomenon — that when a wind of steady velocity occurs at a normal to a cylindrical shell, i.e. such as a tall chimney-stack or distillation column, it results in the formation of eddy currents (Karmen Vortex Trails), which act transversely to the wind direction and which in the case of a cantilever structure such as a tall stack or column, produces transverse oscillations. When the oscillations or deflections reach a critical value, resonance occurs and the deflection then rapidly increases and if not damped out by a change in the wind conditions or by some external agency such as wire *stays* secured to the upper sides of the structure, it will lead to instability collapse.

By increasing the thickness of the shell of the columns, the inertia was increased and this meant that the resonance conditions would then only arise if much higher wind speeds were encountered. It was also established from meteorological data for the construction site concerned, that such wind speeds, capable of generating the resonance conditions in the thicker shell were unlikely to be encountered in the high velocity range, and hence the stress levels would not reach failure conditions.

This approach to the problem provided a design solution based on *stress analysis*. It was innovative but not inventive, i.e. it used conventional design techniques and adapted them to the problem which can now be seen to be the containment of resonance conditions to safe limits by increasing the moment of inertia of the stack.

Some months later I observed that the National Physical Laboratory in the UK had filed patents for a method of eliminating the higher levels of resonance conditions met in *tall stacks*. This consisted of spiral *strakes* which were secured to the upper region of the stack (Fig. 8). They were in effect wind baffles. It was an elegant aerodynamic solution which broke up the Karmen Vortex Trails and hence prevented the deflections amplifying to resonance proportions.

The aerodynamic solution was innovative, but it was also inventive and evidently superior to the method I had employed. A clear case of the innovative response being conditioned by the *innovator's environment* and discipline.

Examples of the *innovator's environment* are found within the kind of project teams employed by many companies in their research organizations and in particular those which are constituted for the express purpose of generating new ventures in the field of manufactured products. Other teams such as those set up to provide technical services support for existing products can also provide an innovative environment but while such teams have excellent market feedback on products and applications they may be too problem solving orientated on maintaining the sales of existing products to open up innovative solutions, which would lead to the generation of new products. This I believe, stems from a desire and determination to keep product specifications changes to a minimum, but in spite of this the opportunity for new product ideas is evident.

We find similar situations in the university research team with greater emphasis on the more fundamental aspects of a product field. Those universities which have set up their Industrial Liaison Groups and have organized one element of their resources around a specific technological and market area are able to offer another field for the *product innovator*.

Figure 8 Wind strakes on a tall stack

Many Government research laboratories and industrial research associations naturally also provide scope for the innovator, but some are primarily concerned with the establishment of design and manufacturing specifications and the preparation of Codes of Practice for the guidance of designers and manufacturers or constructors. The importance of these bodies is that they provide a source of information and test facilities combined with professional expertise in specialized fields which the lone inventor can avail himself of, as well as an environment in which the creative may find employment for their skills in the generation of the codes referred to, by demonstrating what could be attainable in terms of products to meet specific application needs. Such information provides much of the basis for Trade or Government Department, product specifications.

A typical example of the work of such organizations is provided by preparation of design codes for the thermal insulation of buildings etc. by such bodies as the Building Research Laboratory etc. In the Trade Association category we find work on the development of fire barrier upholstery being supported by the development of test equipment and associated design codes. The Furniture Research Laboratory also operates in this field while the Department of Trade in the UK sponsors research by other Government departments or by industry in order to introduce purchase specification codes for their own use and as a preliminary to possible legislation or introduction of National Standards in the UK. The latter being through the British Standards Association.

Outside the categories mentioned so far, we find small specialist research companies which are in effect, self-contained organizations that mirror to some degree the new venture project team of the larger industrial company. Such specialist companies seek to make their sole business the generation of ideas and inventions which they market to manufacturing companies, who they identify as likely to invest their own resources in taking a project forward, usually from the prototype, or development stage to one of design and production.

In complete contrast to all the examples given we turn now to the inventor-entrepreneur who usually works alone, drawing occasionally on other specialists for specific help and information. His environment is initially determined by his previous experience and contacts perhaps gained from working in a research team or in some other corporate but related category. Subsequently in his lone entrepreneurial role, his projects lead him into regular discussion with the potential market for his work and into close liaison with selected suppliers of the raw materials utilized in his development *product*. In a similar way, liaison is also established with suppliers of components, but most important of all with prospective candidates for the acquisition of his Patent Rights and know-how etc. This helps him to establish an environment from which secondary innovation in particular can arise and lays the ground for the eventual exploitation of the project.

3.2 TEAM INNOVATORS AND INVENTOR-ENTREPRENEURS

I had the good fortune to participate in new venture activities in the capacity of an applications specialist supporting several research teams and subsequently in the role of a freelance lone inventor-entrepreneur. This provided me with the opportunity to compare the innovator's role when operating in either environment and to identify the inherent relative strengths and weaknesses which arise in the way they each seek to realize the new product objective.

Turning first of all to research teams; those I was associated with formed a key part of a research department of one division of ICI Plc which was responsible for production and marketing of many different products, some as feedstocks to other industries and some sold direct to the market outlets. Within the division the product range was divided between several business groups who included in their operational responsibilities the encouragement of research for products improvements with which to underpin existing business.

The Research Department, operating under its own budgetary controls also had opportunity to explore some speculative research target areas which it identified as capable of generating new businesses. These were not related of necessity to any existing business operated by the company and were therefore specifically representative of new business ventures whose basis lay in the innovation of new products.

It will be obvious that these speculative research projects needed to 'sell themselves' to a Business Group in order to win their support for taking an idea or invention from feasibility demonstration to an exploitation decision point. Such action would entail embarking on prototype development and applications research for which a semi-technical process plant might be required, entailing considerable capital outlay.

In this environment the team innovator is in much the same position as the freelance inventor-entrepreneur who also has to 'sell' his ideas to those who control investment, or who have the power to influence where investment of development revenue and ultimately capital for process or manufacturing plant, is to be allocated.

The make-up of a New Venture Research Group within a research department is dictated by the category of products targeted on. Typically, research chemists, physicists, materials scientists and engineers of various disciplines formed the core of the teams in the Research Groups of my acquaintance. Business and marketing teams usually work closely with the research team but some companies integrate research, marketing and business management in one organization. It seems to me that the latter is to be preferred in many cases.

The exposed position of the lone inventor will be evident when we compare it with the range and extent of the team innovator's environment in terms of the skills and financial resources available to him. In the areas of technology and marketing, the freelance inventor is faced with many problems as a result of limited resources.

On a more optimistic note, although many difficulties lie in wait for the innovator in general, there are some encouraging aspects for the lone inventor in particular to discover. These include the readiness of many companies to listen to proposals for new products, provide free issue materials needed to facilitate his research, carry out sample product testing for evaluation at their own expense and generally encourage the inventor in the progress of his ideas and invention in relation to their interests.

On the other hand, some companies show a reluctance to recognize the peculiar problems of the lone inventor with respect to his limited resources in terms of financial and specialist skill factors. They regrettably possess a self-constrained view of any new venture, restricting their interest to processes and products with which they are currently familiar.

This latter view, is I feel influenced by a preoccupation with the productivity of their existing processes and a failure to give full weight to the potential for new products to provide *added value* to their existing products, especially where the innovation uses their basic products as starting or feedstock materials.

In the limit the lone inventor cannot expect manufacturers or others to take up his ideas if it is considered that the substance of them represents too great a financial risk for their business, but the value of novelty implied by securing product and process patents is one way of increasing the chances of project success and of gaining its acceptance as a viable development and manufacturing proposition. By comparison the team innovator's role, although similar in many respects, avoids many of these problems as the investment and support for the project is implicit in the corporate objectives of the company or division.

The ability of an innovator to fit in with a team may appear to present us with a paradox as we all recognize that creative individuals tend to 'go-it-alone'. However they are also usually amenable and enthusiastic in their approach, but getting an innovatively inclined individual to follow formalized lines of enquiry or work planning does not usually produce the most effective way of progressing a new venture.

As a case in point; during the period when I was engaged on process plant design, one senior engineering manager said to me, 'the Company doesn't like specials'. He extended this comment with the view that creative engineers are difficult to pigeon-hole or categorize in the library of skills and resources demanded by a design engineering organization and he concluded that they were therefore difficult to manage. I have to agree with the latter comment!

Expressed a little more bluntly these remarks added up to saying that in some large design orientated organizations the characteristics of the innovator can be seen as a barrier to progress in the conventional career structures which place emphasis on conformity and a traditional approach to problem solving. Naturally there are many exceptions to this rule.

A recognition that such attitudes exist among some managements is to dispel any illusion that the innovator's path is easy. This reluctance to capitalize on so-called special skills inherent in the inventive members of their organization represents I believe one of the reasons for the failure of some companies to effect changes in design or product to meet a changing market opportunity, but in fairness to those concerned in my own experience, it became clear that the innovatively inclined personnel should be quickly introduced into a research, rather than a design or production environment.

It will be evident to the reader that the team innovator is not generally recognized as a *professional* in his own right, but must demonstrate his special skills within the framework of one of the professions such as chemist, physicist or engineer etc. In the process of innovation the inventor finds that he must learn to extend his professionalism to other fields, e.g., the engineer, specializing in the innovation of new materials must acquire sufficient appreciation of the chemistry and physical properties of materials to enable him to effectively apply his own discipline of engineering to his project and produce novel ideas and invention in consequence. Happily for the team innovator the opportunity to acquire the relevant knowledge from other disciplines is facilitated by the close integration of effort with other disciplines in the project team.

Such a situation as that described need not discourage the potential team innovator since it is precisely because he finds himself in a schedule work situation that opportunities arise to the creatively inclined for inventive contributions which may relate directly to the task being undertaken, especially where management ensure that he is kept in the centre of the scientific communication flow. Conversely ideas may emerge which lie completely outside the contemporary field of activity, with the consequent possibility that a new field of endeavour or a new venture project may be injected into the team environment.

In the limit, creativity cannot be contained. It makes its own opportunities and creates its own environment whenever it instinctively finds a door is open, or one which can be opened.

Where the self-motivation and creative elements are recognized in an individual team member, enlightened managements give the innovator his head and this often produces projects which are *inventive* and based in the discipline of the innovator.

Clearly the composition of a team undertaking a project is determined by the field in which the idea and product are identifiable. In *new materials* projects, such research teams as already indicated, often consist of scientists drawn from several disciplines including chemists, physicists, materials scientists, engineers, etc. Backing the research team, is the marketing team which plays the highly important role of searching out the market opportunities for the new product and aiding its exploitation by securing the co-operation of suitable potential customers with whom evaluation of the prototype product can be concluded.

Innovators in corporate research organizations also operate in a highly integrated way with their colleagues in the design and marketing teams respectively, while the lone inventor-entrepreneur is faced with a completely different situation since he has not the same degree of cross-fertilization opportunities for his ideas.

Perhaps the most useful conclusion we can draw at this stage of our considerations is that *innovators will always seek to define their own roles*. In consequence, it is not possible to draw up tight job descriptions for them without introducing those elements of constraint which are the antithesis of the inventive temperament.

Team inventors in particular are faced with the difficult task of ensuring that their essential need to maintain involvement with each stage of a project is respected by their professional colleagues, who may see them as a threat to their own autonomy and career ambitions. A sense of humour and a perspective of other people's interests is vital if the satisfactory level of collaboration needed to ensure success, is to be secured in these circumstances.

The inventor-entrepreneur, conversely faces a degree of isolation which denies him the constant stimulus and input of knowledge provided by the specialists usually found in the project team. As a result he must undertake many tasks in the innovative process himself if he cannot obtain the services and skills of others, due to economic or logistic considerations.

Among the tasks facing the lone inventor, which are often in the province of other disciplines, are those of patent application and preparation, market appraisal for the potential product, study of the economics of production and overall business appraisal of the viability of the project. These tasks, are complementary to that of invention and in the case of a corporate project team would be the responsibility of the patent agent, marketeer, research economist and team business manager.

Clearly the lone inventor has his hands full if he lacks the support indicated, but if he is to succeed, the tasks have to be done by him to a level of competance which will enable him to present an exploitation strategy to potential clients for his invention at some appropriate stage. His alternative course is to seek consultancy advice and services which can be found through the innovation centres in the UK and through a variety of small-scale business support organizations set up by local authorities and representatives of industry with governmental support.

The differences existing between team and freelance innovators are eclipsed by the common contribution they can make in the areas of idea generation, inventiveness, development skills and application/market alertness, which form the essential hallmarks of any new venture project.

Any exposition of the team innovator's role would be incomplete if it ignored the wider context in which his task is set. In my own experience the types of project in which I was involved required extensive contacts with manufacturers and research bodies in a number of countries in the Western world. This provided an open window, or more usually, a partly opened window, for observing and learning about the technology related to the field for which the new venture projects were targeted.

The range and variety of projects in which we were involved meant that a number of different industries became the focus of attention in the search for outlets and applications for the product. Among these were the aircraft, automobile, furnishing, ceramic, coating, thermal insulation, civil engineering, and composites industries.

It must be evident to the reader that the extent of these contacts provided not only technological and market information, but an inherently pleasurable experience of travel and collaboration with people of different cultures and outlooks from our own. In fact it contributes to the innovator's stimulus because in such contacts the innovator is continually meeting different ways of identifying and resolving problems of product exploitation and evaluation. Such variety of experience and views provide a fertile ground for innovation.

In parallel with the visits made to explore user needs and open up exploitation opportunities, the contacts made with various research organizations and statutory bodies also proved to be a valuable environment in which to appraise the '*state of the art*' in the project under consideration. This was especially true in the USA where access to the correspondence between the various competitors in the field and the statutory body were available for public scrutiny. An example I had of this was in using the facility afforded by the Federal Aviation Authority in Washington to examine correspondence on *aircraft fire barrier seat design*. This detailed the type of solution to the problem of seat safety which each correspondent advocated and the 'state of art' at the time. The value of this to the innovator need not be stressed further.

To those who might want to make a decision on the direction that they would like their own career to take, the environment of the innovator in the large research organization may now be seen to have special attraction since its pursuit, provides for those whose work is targeted on world markets, both technological job satisfaction and the exuberance of extending one's own horizons in terms of travel, events and people. What more can we ask of our industrial life? In my fortunate case the consequences of vigorously pursuing the exploitation of an idea generally in the form of secondary innovation, as a new venture project, provided as a by-product probably enough visual and philosophical experiences on which to base a travel book as a compendium to the current volume. You may even consider that it might have been better if I had switched my sense of priorities!

In contrast to the team innovator, the lone inventor-entrepreneur will now be recognized as a highly self-motivated individual who must largely depend on his own experience for much of his work, but he will have the satisfaction of exercising the maximum degree of control over his own projects in the initial stages, but be increasingly dependent on others to help him in bringing his ideas and/or inventions to successful exploitation and conclusion.

I suggest that for those who already possess the advantage of industrial research experience, the role of freelance inventor-entrepreneur offers both a challenge and its own special rewards, not least of which is the fullest release of his own creativity.

3.3 PROMOTING INNOVATION

'*Any new venture project needs a champion*' is an oft-quoted phrase, which holds true for the innovator during every phase of a project, but it is at the point that credibility has to be established, i.e. at the end of the feasibility phase that the doubters must be vanquished if the new venture is to go ahead. Identifying a *champion* is not as simple as it might at first appear. The personal career aspirations within the originating company may influence the extent to which individuals are ready to commit themselves in support of a project. It is often safer and easier in some organizations to be to seen to only 'back the winners'. That can mean minimizing risk-taking and can distort judgement and dominate preferences. The *champion* of a project has in fact to share the enthusiasm of the innovator, his confidence in the project — and its risks.

The other particularly critical point where a *champion* is needed is on entering the exploitation phase. Here the *champion* also needs to have financial control over its future, or at least an influence on those who may administer it. Promoting innovation is highly dependent on securing this kind of commitment in the case of both internal and external exploitation strategy.

The lone inventor-entrepreneur has a special problem when it comes to seeking a *champion* for his project, since he will have to gain support from within the company which he has identified as the focus of the exploitation of his proposal concerning ideas or invention. In other words, he has to find the individual within the company concerned, who has the power or influence needed to make the necessary decisions for the adoption of his proposal and he must secure with his help, the confidence of others in the project, its product and himself.

This search for the *champion* of course involves confidence building in two directions, i.e. from innovator to *champion* and vice versa. On occasions I have found that those who appear to support a project, may do so for reasons which only serves an internal and confidential strategy of the company concerned and not the mutually beneficial outcome initially planned and shared.

An example of the latter occurred to me when I was attempting to produce a range of samples which would demonstrate that the effect I had observed could be optimized and lead to a new product. Although I was encouraged to press on with my efforts by the collaborator, it subsequently turned out that the company in question had already carried out studies on materials in a way very closely associated with what I was doing. My own work served to provide a back-up to their own endeavour without me realizing it! Clearly the attempt to secure a *champion* in this case had failed to materialize.

To the freelance inventor-entrepreneur, a misdirection of his energies due to the circumstances described above can have serious consequences to his cash flow. Assessing if a collaborator is genuine in his encouragement and concerned with an equitable outcome is of considerable importance.

Support for a project may have to be sought by the lone inventor from interested Trade Associations or research bodies whose professional standing may serve to make up for the absence of a specific individual *champion* from the market. The commitment of an individual to the value of an idea or invention from such organizations will clearly enhance the chances of gaining acceptance in the market place at a later stage.

It must be obvious that one of the priorities for the innovator is to maintain the confidence of his *champion* as the project progresses. That means keeping him informed of those research and evaluation stages which form the core of the four innovation phases, i.e. feasibility. Applications, development/design and exploitation. Clearly the champion can only maintain his own position of credibility if he is always updated on the critical issues which determine the project's status.

One further possibility for securing a *champion* for a project is the sponsorship of freelance lone inventors by government and industry. Such sponsorship could enable government to ensure that innovation is not over-occupied in improving existing processes and products dictated by managements who are reluctant to extend their energies or thinking to new ideas. It can instead sponsor the pursuit of more fundamentally inventive themes with an emphasis on new products and processes which might otherwise be left to other countries as competitors, simply because in the initial stages some ideas do not always demonstrate a clear probability of success in terms of early profitability.

Sponsorship of the kind described makes a major change in the prospects for the innovation of new products and processes by helping the lone inventor to overcome the cost barrier which brings so many projects to an early and premature end, while increasing the interactive element between the lone inventor-entrepreneur and the research teams of industry, university and other research establishments.

In the UK, innovation centres already seek to build a communication bridge between innovators and the universities. This places its emphasis on supporting business ventures which have their own financial resources and processing or manufacturing capacity, but lack the scientific or technological capability to implement ideas or invention in terms of commercial production. Such centres also seek to introduce the lone inventor to universities or businesses who are known to have an interest in related fields and this can initiate a dialogue on possible financial backing for a new venture in association with collaboration on various aspects of it. These kind of collaborative arrangements are often offered on the basis of the 'backer' acquiring a percentage of any patent/know-how equity property which the project may realize eventually. This is a major concession for the innovator or inventor to make, but alternatives may not be available.

The lone inventor may imagine that he is less in need of the technological type of support, when he approaches the innovation centres at this stage of his project, than he is of financial and practical support. This is at times a misconception as inventors often have the power to visualize a product but lack the knowledge required to implement it, yet clearly financial considerations are often the means of accessing the technological support he may need. Where the required support is gained, it enables him to take his invention to the stage where it is possible to produce a prototype for technical evaluation and establish the means for producing samples or prototypes for market appraisal to support and formulate his exploitation strategy. Once this is achieved, the machinery of government in the UK, may accept his project as a candidate for financial grants to produce the product in sufficient quantity needed to facilitate full-scale evaluation in relation to the market and various statutory or other bodies.

For clarification on the latter point; In the UK the Department of Industry in 1982 published a document entitled *Support for Innovation*. This document makes it clear that the qualification requirements are that applicants for support must already possess the managerial, commercial, financial and technical capability to carry a project through to 'production'. Other requirements will be found in the text of the document.

Some new thinking in the whole field of support and liaison for inventor-entrepreneurs in particular seems to be called for. The area I identify as the one of greatest need is the provision of facilities to aid the inventor at the feasibility phase in particular. It is here that he has often a need for obtaining sound technological advice without risking the loss of his project's confidentiality. Also at this stage he needs to be able to rapidly assess if other products exist on the market that are competitive with his own ideas. In particular he may wish to establish contact with other workers in the same field with a view to a joint venture in order to rapidly bring his ideas to market exploitation.

One highly relevant factor which can facilitate the integration of the individual inventor and the major research organizations in both university and industry is the communications explosion of the present day. The personal computer is ideally placed to facilitate such a revolution in the use of the resources and skills inherent in many creative and professionally trained people who have moved out of their traditional employment in the last few years as the manpower demands of the large companies have changed in the face of world economic and technological factors.

Access to data bases via *modem systems* is already established, but the proposal here is that a rapid and efficient exchange of ideas could be promoted between participants in the field and combined with the provision of facilities for rapid literature and patent searches through this medium. A strategy for the most effective way of furthering specific ideas and/or invention could then be determined.

Clearly the consequences of such a scheme would mean that the time scale, from idea generation to prototype realization could be compressed to a remarkable degree. Conversely, impractical or commercially unattractive ideas could be quickly consigned to the waste-bin or shelved until a new climate favoured their resurrection, but the chief gain would be the stimulus provided to innovation of new products as a consequence of the increased level of awareness gained by the innovator, of work by others in related fields. In fact it would minimize the number of times when 'the wheel is re-invented'.

The cost savings, potentially attainable by a scheme of this kind must be considerable, since a preliminary assessment of an idea can be made by the efficient interchange of information before embarking on extensive experimentation on the assumption that the information accessed will include technical or scientific data which may give positive evidence obtained from related experimentation by others.

The kind of ideas exchange typified in this suggestion could complement the data bank systems such as *Irex*, which is operated by Ideas and Resource

Exchange Plc. This aims to match the interests of those with ideas to those of developers, manufacturers and investors by means of a data bank and computer matching program. It will be recognized that this function differs from the scheme proposed which would aim to introduce direct personal interchange via the computer modem, between individual inventor-entrepreneurs and research teams in industry and university as well as providing data bank access over a range of interests.

Having made a preliminary case for the information exchange, I must hastily add that this in no way is meant to detract from or displace the undoubted importance of experimentation and observation of effects as a vital route for invention, but the kind of preliminary dialogue referred to can at least help to 'target the areas for effective experimental research, while providing an assessment forum'.

The proposal would demand some care in its implementation if adopted, as there is the obvious danger that too much preoccupation with preliminary searches and information exchange may atrophy any thoughts of conducting exploratory practical work to test ideas out, regardless of the observation of others or their failure to do so in the field concerned. On this argument we can see the proposal as one aspect of the innovator's screening process for supporting his search for new ventures.

Continuing the argument for the computer data bank and consultation proposal, I can personally visualize the speed at which such interchanges of ideas could grow providing that adequate safeguards on confidentiality and potential patent property can be established in an uncomplicated format.

The impact on employment, in terms of the participants in such arrangements should be significant, especially if we anticipate that the results would be successful in generating the kind of business venture where early entry into a market with a new product is a major factor in achieving success.

Perhaps the most radical aspect of this proposal is that it envisages a change in the pattern of employment which has already been seen in the USA and to a much less extent in the UK, i.e., by utilizing the skills and knowledge available in individuals who can operate in support of a wide spectrum of industry from their own base at low overhead cost through the versatility of the computer-aided communications system.

In practical terms the proposal would necessitate some form of inventory of participants and data-bases within a subscription system. If developed and implemented it would provide the lone inventor with the resources he would normally find within an innovating company research organization, but without some of the constraints. Evidently it could be adaptable to the corporate innovator also.

The need to promote innovation is closely allied to the demand for novelty in new products, since products generated by the invention route have a distinctive advantage over products which have none. Many industrial nations recognize the need for innovation and support it in various ways but the critical importance of invention within the innovation process is apparently not reflected in that support.

It cannot be over-emphasized that novelty secures invention in terms of patent rights and enhances exploitation advantages. Encouraging the formation of businesses based on novel new products and processes in order to secure an advantage with which to enter the market needs to embrace greater utilization of the inventors in society perhaps in the ways suggested earlier.

The trend to encourage product innovation in general is especially true for those countries with long-established basic industries in which the return on investment, from research and development targeted on existing products and processes has diminished progressively relative to the life of the industry until it has reached the stage where it is uneconomical in many cases to continue with further investment.

We have already commented on some of the types of help available to the inventor but while these forms of aid are commendable, they do not effectively embrace the inventive phase concerned with achieving the prototype product. It may be that this stems from a limited appreciation and understanding of the innovative process itself, particularly where invention is central to that process.

This latter attitude is symptomatic of a remark made to me one day when a marketing manager said 'If you sit down and first work out the market potential for your idea you can save a lot of wasted time and effort spent in experimentation'. While I agree with the sentiment of what was said and have used it in my 'communication proposal' earlier, it has to be noted that the marketing orientated mind does not always see that initial ideas can be transformed in the course of experimental observation and lead to the definition of totally different products than those initially envisaged, perhaps exhibiting properties or effects, which would consequently require a different market to be considered.

This difference in how the innovative process takes place, is tied up with the all important fact that *creativity is often the result of experimentation and what is created determines the market which can be targeted*. It sounds like a restatement of the technology push *v.* market pull philosophy which we have rejected in Chapter 1. In fact it is a restatement of the interactive model of Rothwell, i.e., it is an interactive approach to the interpretation of innovation which finds expression in the innovation project model.

The hub of the argument is that 'if we don't get into the water we will never learn to swim'. Observation and market assessment are highly important but to make an assessment without any experimental feel for the product, process, or effects which are attainable, is to expose a project to a false judgement and *preclude the possibility that the really worthwhile invention awaits the outcome of such exploratory searching*.

Perhaps the most important criticism of the view that evaluation of an idea can be made without any experimental phase is that such statements display a complete failure to understand that *invention is concerned with surprise and not anticipation*. In consequence, it should be evident that trying out ideas, is the very step which is likely to yield invention.

Where invention has already been made, it presents a different situation from the former, i.e. we can proceed immediately with further experimentation in the field of applications, or we can take the initiating idea for the application and first subject it to a market assessment. This can at times save effort from being misdirected and avoid wasteful expenditure on uneconomical applications and exploitation strategies. The weakness in the argument is that in adopting the latter course at the cost of eliminating experimental research, we may again miss the significance of secondary invention which can arise in the application phase.

It is hoped that the proposals made in this section will suggest, to those responsible for promoting innovation, that a review of existing schemes for supporting innovation is needed, with the aim of ensuring that the critical phase of *idea and invention* is given greater backing by government and industry. A positive response would increase the probability of achieving the transition from idea to prototype product and ultimate business success with many new ventures.

Gaining financial support is crucial to any project advance. The inventor-entrepreneur is faced with considerable difficulty, in the UK in particular, when attempting to exploit an invention through collaboration at the development stage with potential manufacturers. Many of these problems relate to understandable caution, on the part of industry and in particular those companies engaged in manufacturing, but it is undeniable that some companies show a marked reluctance to take reasonable risks when faced with a novel product requiring a development input.

The alternative route for the inventor-entrepreneur is to seek venture capital with a view to forming a manufacturing company of his own, but such capital is normally only accessible if it is required for introducing a development product to production. This presupposes that the inventor has funded the invention through the transition from laboratory prototype to development prototype, a stage which is often beyond the scale of the resources at his command! It also begs the question of the ability of the inventor to turn into an efficient business manager. The two roles do not sit easily under one hat.

In complete contrast to the difficulties met by the inventor-entrepreneur, the inventor in the project team of a large company is to some extent relieved of the financial problem of securing funds for his venture by the hoped for backing of his senior management. He also escapes the subsequent burden of managing the design stage of the project, but he faces somewhat different problems which include those of interacting in a constructive way with the marketing organization of the company, an area where fundamentally different objectives and approaches to innovation are often, painfully evident.

One aspect of the team inventor's role is that he may not have budgetary or management control of the project and this inevitably puts him closer in spirit to the lone inventor-entrepreneur, sharing with him the frustration of conflicting philosophies to his own.

Those currently engaged in corporate research in the large company provide a contrasting picture to that of the lone inventor-entrepreneur's role which may encourage some to break out of the constrictions of their current employment structure to undertake the entrepreneur's task themselves and so bring the fruits of their own experience to the creation of new business opportunities in a more direct way. From the foregoing remarks such a step can be taken with at least a general idea of the pitfalls as well as the advantages of the freelance innovator role.

Promoting innovation from within a management structure is a task that many managers of research teams, business, marketing, and design groups etc., can do from their position of autonomy, but much of the pressure to work innovatively must fall on the individual innovator or more specifically the inventor, whose own viewpoint is often the least consulted when formulating departmental strategy.

It is important to remind ourselves at this juncture that when we refer to an 'inventor', we are not indicating some rare breed of individual remote from the areas of professionalism, but to those who already may have a role as scientist, engineer, marketeer, etc. and who also possess an inventive attitude to the problems and opportunities they face in the course of their normal professional function. This is not to preclude that happy band of inventors who without any technical knowledge often bring forward ideas, that can be implemented with existing technology and which may prove to be highly successful in terms of exploitation potential.

Promoting innovation as a philosophy, is worthy of greater attention in the centres of learning where many students are preoccupied with existing curricula and may be unaware of the significant differences between transverse and vertical approaches in product innovation. Perhaps the emphasis on logical analysis by which science is largely taught, needs some reappraisal to ensure the development of the creative potential of students. These comments may help to focus the student's attention on the qualities required for success in an *innovation* type career. Requirements such as a sound professional base in a chosen discipline, interest in the functioning of the marketing element in business, an aptitude for thinking transversely as well as vertically on receipt of a problem, but above all, a desire to create or invent.

3.4 INNOVATION AND MANUFACTURING INDUSTRY

In the United Kingdom, year after year we have seen the demise of some of our manufacturing industry in common with many other countries, while a growth of service industries has occurred. Many voices have spelled out the importance of arresting this trend and it is argued that the creation of new types of manufacturing industry is crucial to economic and probably social success.

The targets for the innovation of new products and processes are obviously numerous, but a prime target lies in areas of technology which are capable of generating a chain reaction of support activity. In this way such innovation can serve the need to create a widening demand for various skills and consequently

Table 1 The emerging pervasive cross-sectoral impact of new technologies (from Gardiner and Rothwell, 1985) (Reproduced by permission)

Sector	Microelectronics	Information technology	Advanced manufacturing	New materials	Biotechnology[†]
Final consumption		*			
Local Government		*			
Transport		*			
Utilities		*			
Metals		*	*		
Rubber and plastics		*		*	
Ships		*		**	
Construction		*		**	
Paper	*		*		**
Food					**
Non-metallic materials	*	**		***	
Mining (land and offshore)	*	*		*	
Textiles	*	*	**	**	
Mechanical engineering	*	***	*		
Business	*	**	*		
Health	**	*		*	**
Machinery	***	****	**	*	
Printing	*	***			
Agriculture	***	*			***
Electrical engineering	*	*	**	*	
Chemicals	***	***	**	*	***
Instruments	**	**	***		
Vehicles	**	***	***	*	
Aerospace (civilian)	*	**	***	**	
R & D	***	***		***	**
Electronics		***	*	***	*
Totals	**29**	**42**	**25**	**25**	**15**

[†] The impact of biotechnology will increase rapidly in the twenty-first century relative to the four other major new technologies. Clinical and field trials and scaling up induce a delay for biotechnology developments.

generate new opportunities for increasing employment. One such target area is in *new materials*. Professor J.D. Birchall has consistently argued, in his lectures, that the emphasis should be on inorganic materials in order to displace organic based products which are eating up the world's fossil resources. As he puts it in his John D. Rose Memorial Lecture (1983b) — 'the history of science and technology leads me to expect future discontinuities; they will probably arise from "irrelevant" investigations'. The surprising introduction of inorganic materials in areas traditionally the province of organic materials, is one of his own examples. My own example is more narrowly focused and forms the subject of Chapter 8, where the point of discontinuity is clearly demonstrated by Fig. 11, which depicts it as secondary innovation which arises from apparently similar types of irrelevant investigations.

Leaving this slight digression for later consideration in depth, the point to be made is that in addition to creating new processes, *new materials innovation* creates *new applications* in which the materials may be exploited. This in turn generates demands for further new manufacturing activities. The argument is a strong case for increased investment and encouragement for research in this field which has only recently led to increases in the funds available for the UK universities, to a more acceptable level.

Gardiner and Rothwell (1985), in Table 1, reveal that while *microelectronics* have generated great impact on a wide range of industries, *new materials* have also performed strongly, but largely geared to research and development and the defence industry. The take-off phase does not yet appear to have occurred with significant impact opposite manufacturing industry.

This delay in penetrating manufacturing industry can be argued to be due to a weakness at the innovation stage of applications research. A reappraisal of the promotion and management of innovation with respect to *applications* appears to be of critical importance. Indeed a fresh look at many existing, and perhaps long established, materials could be expected to yield new applications and stimulate secondary innovation.

As an example of the latter, at one brain-storm session of several research teams drawn from different areas of research and marketing organizations, an existing product was opened up to ideas for generating new business and manufacturing outlets.

About a hundred ideas for product modification were put forward, many of which were concerned with new applications and a few with basic product modification and improvement in the material's characteristics. The implementation of these ideas in this instance proved to be very unproductive, which tends to reinforce the view that product innovation is highly dependent on experimental, and not subjective observations. In spite of this example the exposure of existing products to a *brain-storm session* is a valid way of stimulating ideas that subsequently find expression in the identification and setting up of a research target, out of which can come innovation, incorporating invention.

It could be argued that the ideas generated in the brain-storm session often

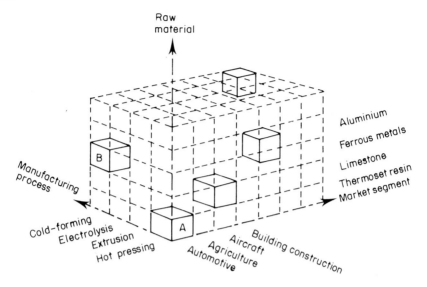

Figure 9 Model of possible marketing opportunities Parker (1980) — *Guidelines to Product Innovation* — Fig. 4. (Reproduced from J.W. Carson, 'Three-dimensional representation of company business and investigational activities', *R & D Management*, Vol. 5, No. 1, 1974) (Reproduced by permission of Basil Blackwell Ltd)

tend to be remote from practicality, whereas the ideas that occur through experiment are already connected to a framework of practical observation and are associated with a liberal sprinkling of transverse observations which could not be premeditated and from which invention often arises.

An alternative to brain-storm sessions as a method of identifying new opportunities for existing products, is advocated by Parker (1980), who cites the use of a three-dimensional analytical model of an existing business as proposed by Carson (1974), see Fig. 9, which highlights the three-dimensional interrelationships that form the basis of the existing outlets for a product. It reveals clearly potential new outlets that have not yet been realized by the product or by some innovated development or application of it.

Parker (1980) discusses many other aspects of the innovative process with a view to alerting management to the critical factors to which they must give priority in terms of resource allocation.

In my view *applications research*, as a distinctive element in achieving a marketable product, needs to be given more prominence in the thinking of all who control a new venture project since it forms the vital link between the basic product or commodity and its use in the market place. Applications research also embraces the need for close co-operation between research and marketing as well as production, where quality control plays such an important part in achieving the specification demands of potential users of the product.

One of the key areas of interest to those looking for support for their innovative efforts, lies in finding financial resources with which to promote their ideas. The funding of entrepreneurial innovation in general in the United Kingdom omits the vital area of the translation of ideas into invention up to the point where prototype definition can be established and a prototype can be produced.

Downstream from this later phase, support for 'prototype evaluation' becomes available but reaching this stage is fraught with mounting costs in terms of securing adequate patent protection, especially in the case of filing applications in several countries and this is compounded by the scale of the exploratory work and sample preparation entailed in arriving at a product definition worthy of advancing to the prototype stage.

From the foregoing it must be evident that much remains to be done to encourage innovation at the emergence of ideas stage in the field of inventions related to manufacturing industry and to 'new materials' in particular.

3.5 MANAGING INNOVATION

It will be evident to the reader that success in the generation of new products and processes, while primarily dependent on those who have ideas and the attitude to them which leads to inventive realization of their possibilities, requires a highly competent team of people possessing the skills needed to turn an invention into a business. The management of new venture projects must therefore have as its prime concern the identification and provision of the particular resources needed to make the transition from invention to the exploitation target as a new business venture.

We have already commented earlier in this chapter on the spectrum of skills that must be deployed for a particular project. Clearly this will vary according to the process and product field. The strategy to be deployed in reaching the product target must be based on consideration of the allowed or available time allocation to the project in relation to the type of innovation involved, the probability of technological success and the expectation of commercial success. Parker (1980) provides a model for this innovation strategy in Table 2, and highlights the main decision areas.

The contribution I wish to make to this analytical look at innovation is that the paramount need in my opinion is for managements to recognize and facilitate the key interactiveness of initial idea, invention, process, market and applications by creating the essential communications and environmental conditions required, in respect of which suggestions and comments have been made earlier in this chapter.

The planning of the new venture project presents a problem if undertaken at too early a stage unless its objective has been clearly defined. Those projects which arise from more speculative experimental enquiry eventually reach the product definition stage which we can identify as the goal of the feasibility phase of the Innovation Project Model, shown by Fig. 4 in Chapter 2.

From this point onwards the extent of planning will depend on the complexity of the project. Parker (1971; 1980) advocated the use of a network which is

Table 2 Product innovation — long and short-term strategies

Available time	Strategy	Characteristics of development	Probability of success	Main decision areas	Likely commercial success if targets met
Short	1	Innovative high risk	Low	Allocation of significant resources	High
	2	Evolutionary low risk	Medium/high	Methods of combating competition	Low
Long	3	Innovative high risk	Medium/high	Degree of acceptable diversification	High
	4	Evolutionary low risk	High	Allocation of resources between fire fighting and new products/processes	Medium

Source: R.C. Parker, *Guidelines to Product Innovation* (British Institute of Management, 1980). (Reproduced by permission)

expressly designed for controlling R & D projects. Following the general structure of the Innovation Project Model Figs. 3, 4, 5, 6 and 7, will in any case be found to provide a practical reference basis for project reviews.

The ideas introduced in the previous section on the integration of inventor-entrepreneurs with research teams of the large companies and organizations, combined with some form of idea/data-bank and computer link, presents management with at least the challenge to explore innovative solutions to managing the innovation of new products.

4 Concept to Feasibility

4.1 THE SOURCE OF INNOVATION

We come now to the heart of the matter — the question of where, when and how, innovation begins. We will have observed in Chapter 1, and in particular from the Innovation Project Model (Fig. 3) that the innovation process starts with the feasibility phase (Fig. 4), broadly as outlined in Chapter 2, where the various key events which take place in establishing the feasibility of producing a new product are introduced. In now considering the source of product innovation in detail, we focus upon the interactive relationships between idea, experiment and invention as the initiating events of the innovative process. The background to our study is provided by two models of innovation, Figs. 1 and 2 (Gardiner and Rothwell, 1985); which are alternative views of the broader interactions encountered between idea, development, design, production and marketing.

To explore the highly interactive nature of the three key events, reference is made to the Venn diagram form of Interactive Model (Fig. 10) which depicts idea, experiment and invention as individual events overlapping each other, designated by areas A, B, C and ABC. Analysis of possible permutations of the three events, reveal that fifteen optional basic situations can provide the starting point for a new venture project. These options comprise any one of three single events, six permutations of pairs of events and six permutations of three events interacting together.

The optional ways of initiating the innovative process are now explored with reference to Fig. 10.

1 Single Event Motivated Innovation

These are the three basic events of idea, experiment and invention, acting independently and optionally to start the innovation chain from a single event source.

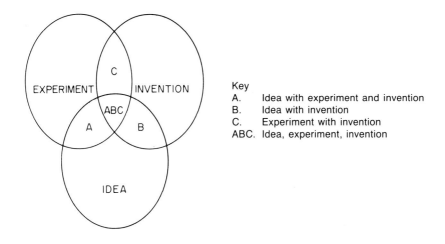

Figure 10 Interactive model. Idea, experiment and invention. Basic interactions

1(a) Idea motivated
On occasions an idea for a new product or effect arises which contains all the information needed to implement it without the necessity for any experimentation. Additionally such ideas may draw upon well established technology and utilize this in a logical and predictable way without generating any novelty, i.e., without introducing invention. Similarly, innovative ideas often have their origin in market need and may have no dependency on invention or experiment for their implementation.

Such products may be 'new products' to the innovator, but since they will use 'prior-art' to produce them they will not be patentable products.

1(b) Experiment motivated
Where experiment is the sole motivator of innovation, it is conceivable that observations will be made which demonstrate that particular 'properties' or 'effects' can be realized by a 'product form' that demands no input of ideas for its usefulness, i.e., its use is self-evident from the experiment. We may also find that the 'properties' or 'effects' have been observed by others and if the way in which we have utilized the existing prior-art on the subject is predictable or logical, we are again denied the possibility of securing a patentable invention, but experiment will have served to introduce us to an existing process and/or business and provided the know-how we need to manufacture our 'new product' in competition with others.

The experiment in such cases will represent the exclusive initiating factor for the innovation chain, but contribute no new idea or invention. What makes such a statement viable is that all innovators at some time re-invent the wheel!

Experiment in this instance will have served our purpose in seeking new venture opportunities by revealing how we can replicate an existing product, already manufactured by others. If the manufacturing process details have been kept confidential, instead of patenting, we shall not be restricted from manufacturing the product ourselves, based on the information gained by experiment. Conversely if we find that the product or process have been patented we will be infringing the patent, if we go ahead and start manufacture ourselves.

The foregoing is not to deny the possibility that ideas and invention may subsequently arise in the course of developing and designing our own process, as the innovation proceeds to its exploitation target, but it serves to illustrate that experiment alone can be the initiator of the innovation chain even where novelty is not secured. In fact, many new ventures are launched on this basis, but as suggested, the result of persistence in progressing the project to a prototype product by means of laboratory experimentation and processing, often pays dividends by yielding sufficient novelty on which to secure an invention claim based on some novel aspect which we introduce subsequently.

1(c) Invention motivated

We have already appreciated that invention may present itself to us without reference to an idea as its source and before any experimentation has been undertaken. In principle we are saying that the inventive act formulates a novel product which is not dependent on first contemplating an idea to which we wish to give effect. Neither has it required the observation of an experiment to yield novelty to us which would constitute invention.

In consequence the invention itself is the sole initiator of the project. In a similar way to that described in the previous paragraph we can expect the project to yield up ideas and demand experimentation in its evolution, but in this particular case: i.e. at the inception stage it is the invention alone which has provided the motivation and initiation of the innovative process which may rapidly result in the introduction of subsequent ideas for the utility of the invention and for research from which further innovation can result.

2 Paired Event Motivated Innovation

We now consider two of the three basic events taking place in such a way that one is continually dependent on the other for its own motivation and development while the third event is dormant, i.e., the start of the innovative chain is determined by the synergy between the pairs of events. Reference to Fig. 10 and the overlap areas of A, B, and C, helps us to interpret what options can occur. Six permutations are available to us.

Area A indicates

2(a) Idea is lapped by experiment but not by invention, indicating that idea in this instance conjecture's on what field may yield a new product and specifies the experimental programme to explore this field, which then feeds back information to reformulate the initial ideas, but in this example it does not realize invention, i.e., it does not identify novelty. The synergy of idea and experiment provide the product and process realization and starts the innovation chain.

2(b) Conversely, experiment, may give rise to an idea for a new product or effect as a direct result of observation. Again, in the absence of novelty no invention will be achieved, but the know-how of producing the product will have been established and merit the pursuit of further experiment, or conclude with a new product definition in which invention is absent.

Area B shows

2(c) Idea is lapped by invention but not by experiment, from which we can conclude that the idea contains the substance of invention in it without recourse to experiment, i.e., ideas for a new product are developed to a point where they trigger their own implementation in a practical way, as invention. The inventive concept is recycled to effect a revision of the initial data and this form of recycling may continue several times as first idea postulates and invention interprets until the cycle concludes with an acceptable new product.

2(d) An alternative possibility is that invention provides, as part of the concept, an idea for its own development and application. As in the previous case this idea can recycle to precipitate a secondary version of the invention to accommodate it.

In both of these instances we have a clear case of innovation which realizes invention without experimentation.

Area C indicates

2(e) Experiment lapped by invention but not by idea. In this case we can conclude that invention arises directly from experiment, but the invention may dictate further experimentation to resolve the questions of how limitations may be overcome.

2(f) Conversely, invention may initiate experiment in order to verify its feasibility, but again the results of this may result in an entirely different concept of the invention.

3 Motivation by Combinations of Three Basic Events

The representation of the combination of A, B and C is identifiable as area ABC. It implies that all three events act continuously on each other but as in the case of paired events, the sequence in which this interaction takes place can take several forms. The number of different permutations basic to our analysis is again six:

Idea dominant

3(a) The most obvious combination signifies that idea may precipitate experiment which yields invention.

3(b) Alternatively, idea may give rise directly to invention which then requires verification by experiment.

Invention dominant
3(c) Invention may be the initiating factor, which then precipitates experiment followed by ideas for product applications.

3(d) Conversely, invention may yield ideas for product application which then demands experimental verification.

Experiment dominant
3(e) The last set of possibilities for ABC is where experimentation has been initiated for reasons which need not be specified, but from which ideas for a product emerge and then direct the course of further experiment which eventually yields invention.

3(f) The converse may occur, of experiments giving rise directly to invention and the subsequent precipitation of ideas for applications.

Summarizing

The important point to make is that where two or three events contribute to the inception of the innovation chain, they interact on each other in a synergistic way, often by a recycling process. It is this dynamic characteristic that makes it difficult to identify exactly how, when and by whom, an invention has been made, or what gives rise to an innovative project which produces a new product. The kind of analysis I have attempted in relation to Fig. 10, will I hope prove to be enlightening to all interested in the intriguing way in which new ventures arise, but I feel it will especially serve those who innovate and invent as well as proving to be of some value to the patent agent when assessing the authenticity of invention claims etc. Last but not least I offer it to the marketeers in the hope that it may help to clarify for them what the innovators are up to in their apparently unpredictable efforts to start the innovative process for the creation of a new product.

4.2 IDEAS AS THE STARTING POINT

While many 'brilliant ideas' are conceived, the facility for translating them into practice may take decades of evolutionary progress in the field of technology. Only when an idea finds the complementary knowledge for its inception can we expect product innovation to get off the ground. The ideas which therefore concern us in the quest of product innovation are those which are often related to the field of our own specialization, i.e. 'knowledge-based ideas'.

It is precisely knowledge-based ideas which can fully exploit even existing products by maximizing product demand. This entails generating secondary outlets for the product through the implementation of ideas often demanding invention. In addition ideas are needed to provide a basis for the generation of new venture projects with which to expand the existing business into more cost/effective 'added value' product areas. Perhaps the interactive model (Fig. 10), will reinforce the emphasis on the importance of the idea as the starting point.

In the light of Section 4.1, we saw that idea and invention are quite distinct from each other although highly interactive. We can now address ourselves more fully to ideas as prime movers of invention of the basic product and as a motivator of a secondary invention to establish applications for the basic product, often in a secondary form.

Ideas are conceptual — they postulate the possibility of a product, effect or application — if specific means of implementing the idea are forthcoming. Such means may already exist as prior-art or technology, but where it does not, invention becomes an opportunity for not only solving the practicality of the idea, but of securing novelty, on which basis exploitation chances of success for the project are increased.

The cases where idea is the prime mover of innovation can be readily discerned in Fig. 10 Section 4.1, and it is useful to apply the analysis to a case history.

The example I have in mind occurred in my early days as an inventor-entrepreneur and consultant when I was asked to resolve a problem associated with wall cavity thermal insulation of buildings. It was a clear case of 'market pull', i.e. the need was recognized by the market, which consisted of the Trade Association, representing contractors engaged in carrying out the thermal insulation of buildings and in particular of houses.

In responding to this request, from the 'case history' which follows, it will be seen how the invention, although relatively simple in concept, together with experiment interacted with the initial idea in a way that fits the interactive model (Fig. 10), 3(b), i.e., idea giving rise to invention which is shown to be feasible by experiment, but which also demonstrated the need for amendment of the initial idea and modification of the invention. A case of the three events interacting on each other.

The points arising in the case history, which serve to show how the innovation process developed, are as follows:

Cavity wall insulation systems utilize a variety of thermal insulating materials, but in this case the problem relates to the use of expanded beads of foamed polystyrene which are blown into wall cavities through holes drilled in the wall.

When beads fill the wall cavity they also escape into the building roof space, or into communicating cavities of adjoining buildings, e.g., semi-detached houses may have communicating cavities, unless the cavities have been sealed at strategic locations.

The method used by the 'trade' to seal cavities was to cap them with insulation slab etc., but this meant gaining access to the roof space or attic, which was very time consuming and presented a risk of damaging ceilings etc. An alternative approach was desired which could overcome the existing drawbacks of the method employed and also provide a cost reduction on the installation.

From this information the idea was proposed by the client that some kind of seal might be inserted into the cavities through the holes prepared for foam filling operations. The problem for the innovator was how to implement the idea, i.e. *the project could not be initiated by the single event of the idea without recourse to invention and/or experiment or prior-art knowledge of cavity sealing.*

Some transverse thinking came into play at this point. I had been concerning myself with thoughts of another project utilizing the resilient properties of alumina fibre mats used under compression conditions.

One of the features of this work was the use of a containing envelope made from a non-woven fabric which held the compressed fibre mat under compressive strain, reducing its thickness. This facilitated entry of the mat into a cavity. Subsequent application of heat during service destroyed the envelope without damage to the fibre mat, which was made of alumina fibres, which then expanded to fill the cavity and produce a 'seal'.

The inventive step for the wall cavity system was now made, based on the initial idea and the background technological experience and know-how from related project work. It took the form of postulating an envelope of polythene film in the form of a long tube, inside which was packed flexible polyurethane foam. The tube was heat sealed and evacuated, then sealed off at the vacuum aperture.

The product could now be defined as a precompressed polyurethane foam sealing strip, enclosed in a plastic film envelope by vacuum packaging. The seal was to be positioned in the wall cavity and then expanded into place by releasing the vacuum in the envelope. This latter step required that the size or diameter of the envelope when filled with the polyurethane foam prior to evacuation, was greater than the width of the cavity it was required to fill.

It might be thought that an obvious solution to the problem already existed in the form of the alumina fibre seal referred to earlier, but this was not appropriate since this seal was far too costly for this application and in fact would have represented overdesign, as the high strength and thermal stability of the fibres would not be required in a static installation such as a building cavity. Another ojection was that the compaction/expansion ratio of the fibre was much less than could be provided by the polyurethane foam which could more adequately accommodate the large volume change demanded by the relative dimensions of the wall cavity and the access hole used for insertion of the seal.

The system chosen depended on the residual compressive strain in the polyurethane foam, to provide the retentive force needed to secure the seal in place, after release of the vacuum in the containing envelope or tube. Clearly other resilient materials, perhaps with inherently superior fire barrier propeties, could provide options to the designer when specifying the seal construction but as pointed out above, questions of overdesign or cost-effectiveness have to be considered.

Experiments with the seal tube showed that it was not possible to depend on its own rigidity in positioning it, *in situ* and that it was necessary to support it temporarily by means of a simple mandril in the form of a gutter section. Once in position, all that was required was to break the vacuum in the envelope by piercing the polythene which resulted in the expansion of the contained foam to effect a continuous seal in the cavity, at which stage the mandril could be withdrawn from the cavity.

This example shows clearly how an initial idea is implemented in terms of invention and verified by experiment, i.e. para (3b) Section 4.1, to achieve a novel solution to a problem of market origin. In this case the initial idea was dominant in the motivation for an innovative solution to the problem, but we need to note that recycling of information and extension of experiment was taking place which resulted in the introduction of the mandril, to make the system a practical proposition.

Case histories can be produced for all the combinations of idea, invention and experiment referred to in Section 4.1 some of which will emerge as the various phases of the innovative chain are considered in succeeding chapters.

Success in projects of the type described, demand more than the ability to think innovatively. A considerable amount of the effect entailed for the inventor is concerned with the techno-commercial aspects of the project, but success depends on being able to cope with the inevitable sequence of attainment and failure as the realization of an initial idea is grappled with in inventive terms. Often elation and disappointment follow each other with great rapidity as problems are encountered, then hopefully resolved.

To address ourselves more specifically to this distinction between idea and invention we need to appreciate that ideas cannot be patented. Inventions can. *It now becomes clear that the distinction is of much more than academic importance for those who seek to make a business venture out of new ideas and invention.*

Typically an idea sets targets for invention, e.g., we may identify the obvious benefit of an upholstery flexible foam which did not emit significant smoke in the event of it catching fire and suggest that some kind of post-treatment of conventional polyester-polyurethane foam could be used to provide the required properties. The suggestion constitutes an 'idea'. Invention is required to resolve the means of implementing the idea and to demonstrate that a specific post-treatment process and formulation is effective. Of course, a number of different innovators may use the single idea to realize a variety of inventive solutions, each of which may possess a unique element in them and so qualify for patenting. In such situations each innovator is faced with competing for the market commanded by the initial idea on a cost/effect basis.

In other examples we find that ideas can also follow invention; i.e., the idea is a reaction to an invention often in terms of product improvement or *secondary invention*, but in my experience, this chiefly arises in the field of applications.

The important point already advanced regarding several ways of implementing an idea can now be re-emphasized. This idea may already have been published by other workers in the field, but the most effective way of achieving the effects desired may still await invention. In other words an existing undeveloped idea can be turned into a business opportunity as a new product, for which patent protection can be secured, if the innovator can bring forward an inventive solution to the implementation and realization of the idea.

The sources of ideas are as diverse as the ideas themselves. In the case of the area of products we are concentrating on, these can arise from the market place or the technical or scientific base, but we must remind ourselves that the commitment of this book is to implementation as manufactured products.

We have already referred to the marketing element in the innovation process as 'market pull' i.e., those ideas which arise from the potential users for new products or effects for which they have already identified a specific potential market. An example is where a manufacturer identifies a market need but finds he has a gap in his product range and cannot therefore meet the need. As a result he may be able to provide a specification for a product to meet the need but not the know-how to achieve it or conceive how it may be realized.

Such ideas for new products may come from marketeers engaged in exploring market needs for related products, but are even more likely to emerge from those frustrated in their use of inadequate existing products. The case history of the cavity wall seal, given earlier, is an example of this.

The housewife is another case in point. She may quickly realize the need for some item of equipment that could iron her washing quickly, and conceive that some particular variation on the iron and the laundry press is what she needs for her range of linen, but she will be less likely to articulate the thought or idea into a concept which produces invention because, while appreciating the market need, she probably lacks the technological knowledge to which she can relate or interact and enable her creativity to find further expression.

The housewife is not alone in her problem. The professional engineer, trained in mechanical engineering finds the same difficulty in articulating the implementation of an idea relating to areas such as electronics, if he has no special skills or knowledge in this field.

This limitation on transferring idea into invention lies in the absence of relevant technical knowledge and may be further inhibited by a lack of ability or aptitude to think horizontally or transversely. It also highlights the importance of finding ways of facilitating the interaction of ideas with those of inventive inclinations and who possess the requisite technological skills and knowledge to implement them.

In the research project teams of some companies, an area where some improvement can be gained for example, is by increasing the involvement of innovative members of such teams with the marketeers in visits to the potential customers in the market place. This approach increases the probability that 'new product ideas' are identified, even where 'market contacts' have not seen for themselves that new applications for existing products are evident which can fill a gap in their current marketing portfolio.

No doubt research and marketing management will counter this suggestion with the view that on a new venture project the marketeer and inventor often team-up on such forays in the market place, but my comment is not addressed to this situation but to the search for new opportunities for existing products, i.e., not new products. The case for this proposal, is that the marketeer is necessarily preoccupied with his own sales portfolio of products which are already in production, whereas the innovator is alert to the stimulus of learning new technology and product specification limitations which ultimately characterize the market need. This inevitably leads to the inception of applications research which adapts existing products to meet newly identified market needs.

This product improvement approach to innovation is the fertile ground for the innovator to not only generate ideas for the applications for existing products but to recognize from his familiarization with product and market what experimentation or other steps might be undertaken to achieve a new product which may be the inception of a new project for some entirely different business area.

Applications research is particularly involved with the potential market and forms the ideal ground for building up the co-operation of innovator and marketeer. This is conducive to the search for new products which depend greatly on a transverse interaction between technological applications potential of a basic material and the market requirement in which utilization may be feasible if the fit is recognized. The transverse element involved is in effect the 'operator' in many projects and the means by which ideas are spawned or invention stimulated.

4.3 THE INVENTIVE STEP

Invention is a highly unpredictable event, yet the same transverse thought processes that lead to the emergence of a creative idea also results in the creation of a means for implementing ideas in practical terms. In other words — they precipitate invention. As we have seen earlier it is not always the case that idea precedes invention, nor that invention always arises from experimental observation. The cases analysed in Section 4.1 provide ample illustration of this.

In spite of the complexities of how invention arises, the task of innovation in which it can play such a key part is not so traumatic as might be first thought. In fact to the creatively inclined it all becomes a mixture of fun, challenge and most of all a sense of fulfilment in response to an indefinable driving force that carries much enjoyment with it.

Much as the 'creatively disposed' find delight in the inventive step, it has to be emphasized at the risk of repetition, that invention may not be essential to securing the innovation of new products in all cases; i.e. many new products do not incorporate the element of novelty which characterizes invention, as Section 4.1 makes clear.

Invention is defined as devising that which is new but more specifically it must embrace novelty. This is most easily recognized in the case of the mechanistic type of device, but can be more obscure when entering into the territory of processing or manufacturing operations.

We learn that invention is only patentable if it is couched in a practical form. This must enable someone who knows the technology to make the product, operate the process or put into effect by practical means, the substance of the invention (HM Patent Office, 1982). Abstract ideas are not candidates for patents! *The significant point is that ideas say 'IF', while invention says 'HOW'.*

It is important to expand on the statement that innovation can proceed without invention and the reader will readily recognize that many items present themselves as 'new' which are in fact only alternative versions of an original product; e.g., a knife with a different shape from the one on your table is not an inventive knife because of its change in shape unless it incorporates some new feature in its design which is novel and surprising. Many such variations which lack the latter, simply constitute design changes, not invention, but if the changes concern appearance and not function, providing they are a 'new design', they can be protected from copying by others by registering as 'industrial designs' (HM Patent Office, 1985).

To reinforce the importance of the design aspect, Gardiner and Rothwell (1985), point out that innovation can be market, technological or design based, but the definition of 'design' differs here from that of the previous paragraph, i.e., the former refers to appearance as expressive of design, whereas in the latter we imply all the functional aspects of a product as well as the associated appearance features. These comments may serve to guide the innovator in his approach to the question of invention and/or design in relation to identifying the appropriate form of protection he needs for his product, preferably by reference to the two Patent Office publications stated in the preceding paragraphs.

An example of a product which can be a candidate for both patenting and design registration is found in such items as a ceiling tile, where the novel formulation and processing developed through the facility afforded by functional design must be protected by patenting, while the numerous variations in surface texture etc., can be registered as 'non-functional designs' within the definition of the UK Registered Designs Act 1949–1961. *In our subsequent use of the term, 'design', will be related to functional applications.*

To provide further illustration of the distinction between idea, invention and design I have selected an example which at the same time sketches out the broad outline of the concept to feasibility phase of the innovative process, following the innovation model itself (Fig. 4).

The example I have chosen for this purpose is an organic/inorganic composite fire barrier structural foam, UK Patent Application 8118638 (Bradbury *et al.*, 1981).

What we are considering in this example is really two distinct projects, where the second project is the result of a transverse shift in the innovative chain

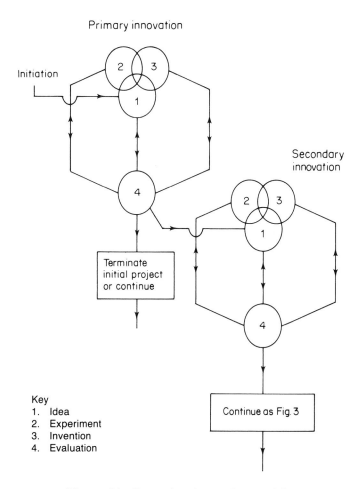

Figure 11 Secondary innovation model

which starts a secondary chain in which the ultimate new product is created. When we get down to the project models all becomes clear!

The initial project had as its objective, which was set by management, the development of a fire barrier spray-up system employing inorganic foam made from layer minerals. The secondary project which it ultimately yielded was one which resulted in the invention of a rigid polymer composite foam incorporating the layer mineral as a reticular foam matrix.

To comprehend the sequence of events and interpret them in relation to the innovation project model it is necessary to appreciate the background to this work.

The inorganic foam was the primary new venture product and it was considered that one of its potential applications was as a fire barrier for structural steelwork in the form of a spray-up system. While the initial product target was quite removed from the ultimate new product definition of a 'composite structural thermal insulation foam', a common factor existed, which was the use of layer minerals possessing important fire barrier properties.

Backing up this common factor was a specialization knowledge of the properties and processability of both inorganic and organic materials associated with fire barrier and/or structural foam applications.

The events in relation to the innovation project model (Fig. 3.), and in particular Fig. 10, have been interpreted for the sake of clarity in Fig. 11, which I have designated as the *secondary innovation model*, since it shows the inception of the final product as a result of a transverse 'shift' in the innovation chain in response to the experimental and evaluation results obtained from the initial project.

Stage 1. The Initial Inorganic Foam Spray-up Project

Following the logic of Fig. 3 to describe the initial project, the events concerned are:

Event 1. Idea
The basic project objective of the research team was the research and development of an inorganic fire barrier structural foam produced as rigid slab or boards. To extend the potential market for the product an applications research target was set by management which was to: develop a spray-up version of the inorganic foam for the fire protection of structural steelwork. This represented a clear example of innovation being promoted by management decisions which also carried with it an idea for a new product.

The first step was to give preference to the concept of spraying dry individual inorganic foam beads or prills in conjunction with an aqueous dispersion of the layer mineral to act as a binder to secure beads to each other and to any sprayed surface. In this way it was hoped to build up the required thickness of barrier foam without introducing the shrink-cracking problems of laying-down wet foam.

Earlier work with the drying of the inorganic foam in slab moulds had encountered severe shrinkage problems but this had been overcome by preforming of the foam into small beads which were then bonded together in a press mould.

Event 2. Experiment
Lack of supplies of inorganic beads and problems on quality control led to the selection of an organic foam bead, i.e. polystyrene, to serve as a model for research purposes. The beads were coated with the inorganic material to provide a test model for bonding. Bonding of the coated beads to each other was shown to be possible, by compacting the beads together in a mould while the coatings were wet, then subsequently drying them.

The bond strength between the inorganic coatings was demonstrated to be weak but it modelled the result to be expected if an aqueous spray-up system for inorganic beads was to be developed.

Event 3. Invention
Curiosity took over at this point. The coatings formed by the layer minerals had been employed by others to provide a flame resistance to materials such as glass-fibre 'papers' and aluminium sheet.

It was thought that the coatings might prove to be effective as a novel fire barrier for the polystyrene beads in the bonded structure and that they could conceivably prevent the kind of thermal collapse which was known to be inevitable with conventional foamed polystyrene structures in which melting and dripping of the polymer constitutes a serious 'spread of flame' hazard.

The first identifiable invention was thus a laboratory process, consisting of a method for applying the inorganic layer mineral coating to polystyrene beads and bonding the beads together by bringing their wet coatings into contact with each other to form a novel 'composite' with potential fire barrier properties. Both process and product were novel and hence formed the basis for what turned out to be a preliminary invention, from which the secondary invention was to emerge. No evaluation of the fire barrier properties had been made at this stage.

Event 4. Evaluation
A flame test on the 'initial agglomerate' was the next logical step and the result was that the inorganic coatings survived and retained their structural integrity, while the polystyrene degraded without dripping. The initial invention had now been evaluated in terms of demonstrating the effectiveness of the two materials when combined in the way described. *It is interesting to note that this invention was not the aim of the targeted research but a consequence of it.*

Examination of Fig. 10 shows that the innovation follows that defined by para 3(a) in Section 4.1, i.e., idea followed by experiment, which on recycling the results, extended the initial idea to envisage the product as having fire barrier properties to produce a rigid fire barrier thermal insulating product. This represented the inventive step, which was then evaluated by flame testing.

Stage 2. 'New Composite' Project

At this point the discontinuity occurred in the innovative process which generated a new project, process and product, through secondary invention. This represented a transverse shift in the innovation model, i.e., from evaluation of the initial 'model material', to the 'idea' for the 'new product'. Fig. 11 shows the displacement of the innovation path which results.

Event 1. Idea
The idea was that a product incorporating adequate structural integrity for use as a structural foam might be possible, combined with a novel fire barrier performance, if the organic component could be polymer bonded while the

inorganic component was distributed in the matrix of the composite to provide the fire barrier properties.

The concept was based on the recognition that the strength of a polymer bond was much greater than was attainable with an aqueous inorganic layer mineral in which the bond strength was dependent upon the weaker van der Waals' and hydrogen bonding forces. Additionally it was noted that the fire barrier properties of the inorganic film, when applied as a coating on polystyrene-foamed bead provided valuable structural stability to the initial 'composite'.

Event 2. Invention
Transforming this second idea into invention of the 'composite' amounted to recognizing that the existing process for making polystyrene structural insulating foam boards, which have very low porosity, should be capable of modification to yield a porous structure which could be post-treated by impregnation with the inorganic material to produce the new product.

The commercial process for making conventional foamed polystyrene insulating boards is by partially pre-expanding polystyrene granules to produce foamed beads by the application of steam heating and then fully expanding them in a mould to bond them together in the shape of a 'highly impermeable structure'.

For the purpose of implementing the idea for the new product it was essential to minimize the second stage expansion of the beads during bonding, i.e. a process control was required which would ensure polymer bonding and at the same time *provide an interstitial porosity* in the polystyrene boards. Clearly this depended on arresting the heating expansion phase in the process by conventional methods of cooling at the critical stage, which occurs after the bonding together of the pre-expanded beads had been achieved, but before the interstitial porosity between them was lost.

It was concluded that if the proposed organic phase process could be achieved, the inorganic material could then be introduced into the interstices of the matrix to produce a 'composite structure' which would meet the objectives of the 'second idea'.

It is important to recognize that in this example of invention, it was made after the 'second idea' and before further experimentation had demonstrated feasibility, i.e. the method of translating the idea into practice was visualized in principle prior to verification by experiment. The sequence of events is as described by para 3(b) of Section 4.1 and as portrayed in Fig. 10.

Event 3. Experiment
The sequence of this secondary invention was different from the first stage. In this case the idea postulated what was needed, i.e., a major improvement in the bond strength of the composite which rapidly produced the inventive step of limiting the expansion of the beads, polymer bonding them, instead of 'glueing' with the layer mineral, then introducing the layer mineral into the interstices which had been established in the matrix. The point is that the

Figure 12 Polystyrene/layer mineral 'composite foam'. Flame test sample — Microphoto (20:1) (Reproduced by permission of ICI Plc)

invention was made without first making recourse to experiment. Experiment was of course subsequently essential in determining the optimum process parameters in order to achieve the product and in demonstrating feasibility.

A highly relevant factor in realizing this product was that close contact had been established with the polystyrene manufacturers and this had provided an intimate knowledge of the processing methods they employed. Augmenting this with a literature study meant that the concept of the final product was laid down in the ideas stage.

Event 4. Evaluation
Laboratory techniques were evolved which produced the envisaged porous structure, i.e. the expansion of the polystyrene beads was restricted after achieving an organic bond by controlling the process steam parameters of temperature and pressure in a mould.

The inorganic material was now introduced into the pores of the organic matrix as an aqueous slurry and dried.

Evaluation of bond strength and fire behaviour showed that a potentially useful 'composite structural foam' based on organic materials, e.g. polystyrene, could be achieved when combined in a specific way with layer minerals of specific form.

The effect of flame on the test samples revealed quite remarkable results as

Fig. 12 shows. In particular it is possible to distinguish from the photograph of the residual inorganic skeleton of the composite, the location, form and size of the original area of polymer bonding which was achieved in the process and which is indicated as holes in the inorganic 'shells' of the original layer mineral bead coatings in the composite matrix, i.e. what we see is a perfect replication of the form and geometry of the polystyrene bead, on which the inorganic layer mineral was deposited, to form a homogeneous fire barrier film.

This type of evaluation shows clearly that the initial visualization of the product is fully endorsed by the subsequent evaluation experiments which themselves contain inventive elements with respect to the 'laboratory processing' of the controlled polymer bonding.

Event 5. Product definition

Expressed in conceptual terms, what had been achieved was a product sample which consisted of a reticulated composite foam structure in which the inorganic material coated the internal faces of the innterstices of the porous polystyrene foam. This characterized the second invention.

The concept of limiting bead expansion was subsequently shown to be capable of adoption to processing on existing polystyrene structural foam-making plant. This immediately provided the project with a potential for full-scale commercial implementation.

Event 6. Patenting

At this stage of the project a patent application has to be considered, but the ramifications of taking out a patent needs to be set against the option of retaining confidentiality and know-how. These aspects are discussed in Chapter 7.

Event 7. Market/cost studies

Market research showed that the elimination of the 'drip' hazard of polystyrene foams would be expected to increase the range of applications for polystyrene in the form of the new composite, since the retention of structure during fire conditions implies that spread of fire is reduced and smoke emissions may also be minimized.

At this stage, of the project, cost estimates of the product could only be notional, since no work has been done in the direction of a commercial scale process. In spite of this it is usually possible to form some idea of the possible competitiveness of the 'product' by considering the material's costs and the major processing cost; e.g., in the type of product being discussed this is the drying cost.

From the data assembled the marketeer and innovator can arrive at some idea of potential market size for a developed product. This will be very approximate at this stage, but it will serve to guide those who must take the decision to extend the research effort to the next phase; i.e., the applications phase.

Event 8. Project review
At this point in the innovation process it is essential to take stock of what has been achieved in terms of scientific or technological results. This constitutes the feasibility assessment against which we must set the cost data and any available market intelligence. In particular the strength of any patent position or know-how must now be taken into account.

What emerges from the review or arises as a consequence of it, is a need to examine the exploitation of the invention through applications to which the market is expected to respond and to consolidate the work done in the feasibility phase. In more direct terms; check the results! Examine the cost predictions and explore the weaknesses etc.

In the project described, the case for proceeding to the consolidation of the invention and tailoring the product to meet applications was approved and its immediate consequence was to undertake more thorough evaluation against the specification standards which must be satisfied if the product is to enter specific markets.

This concluded Phase 1, of the innovation, which has taken us to the establishment of the feasibility point in the project and as far as we need to go with this concentration on the way innovation gets off the ground. The remaining phases of the project were to follow the general pattern illustrated by other examples, including scale-up and prototype production, but of special interest to us here is that this particular case has revealed that sophistication of the innovation project model is necessary for us to interpret the incidence of such matters as secondary invention.

On a more practical level I hope the example has aided our efforts to comprehend how, when and where innovation starts and goes on to adjust to transverse influences before realizing its ultimate product definition and new venture status.

The progressive nature of the innovative process is evident in the two stages of both idea and invention, which took place in finally characterizing the prototype product for a designated range of applications.

4.4 EXPERIMENTATION

In the course of illustrating the distinctiveness of idea and invention we will have observed the importance of experiment.

One point that was made was that innovation can often be motivated on the strength of an idea without the necessity for experiment. Even where this is the case, we find that in order to evaluate the idea, experiment is often desirable before committing resources to the project on a significant scale. The experimentation referred to is of course one of verifying that the substance of the idea is practical, but it is also concerned with establishing the substance of the idea in terms of properties and effects.

Overriding the relatively pragmatic merit of experiment described above, the real value of experiment is that it provides the opportunity to discover or observe some effect or information that a logical analysis of an initiating idea would not reveal, i.e. information or effects that were not self-evident in the initial idea. In my view the reason for this, is that invention is not precipitated by logical progression but as we have already intimated on earlier pages, by *transverse creative intrusions of thought that break the train of logic*. This is why we characterize inventions as possessing 'surprise'. It is also why the resulting 'surprise' is often novel in some way; i.e. in concept, form, process, operation, effect, or application etc.

To take as an example our last illustration drawn from the 'composite material'; we saw that after starting out on a project with a designated idea and product target, to our surprise, we ended up with a different product for a different application. All of this occurred because experimentation was initiated by the first project.

The point here is that the experimentation was focused on a target divorced from our eventual product, yet because the field of interest was common to fire barrier performance materials, it resulted in furnishing ideas which probably would not have occurred to us in any other way and which produced a horizontal shift in the progression of the innovative process. This 'shift' marks the origin of both primary and secondary invention.

4.5 PRODUCT EVALUATION AND DEFINITION

The evaluation stage is of critical importance to the assessment of the feasibility of the original idea which may have led the project or arisen in the course of its progress.

Evaluation consists primarily of asking the question; What are the characteristics of the product which has been realized? In the case of 'new materials', it is evident that all kinds of physical properties are of interest. Typically these are such properties as; mechanical strength, porosity, permeability, fracture toughness, density, indentation hardness, resilience, fatigue strength, dielectric characteristics, thermal conductivity, etc.

Parallel to the physical properties in importance is the fabrication potential of the product. This raises questions of; can it be cut, drilled, bent, or joined in any way?

To arbitrarily evaluate all these properties for every new material is obviously uneconomical. Clearly only the evaluation of those properties that relate to the product's field of use seem to be justified. While this is true in the first phase of evaluation, it would be folly not to subsequently extend the measurements further, as a surprise may await the investigator that could yield important new applications and secondary invention.

In selecting the properties and manipulative or fabrication parameters for measurement and assessment it will be evident that the potential use or applications for the product will play a critical role in determining these.

In the early stages of a project we are not necessarily aware of what these potential applications may be and for this reason the characterization of the product in definition terms must be kept fluid until experience gained from market contact and further experimentation has had the chance to shape and identify the parameters which give the product its distinctive character.

These parameters may prove to be novel 'effects' which are the result of utilizing the properties of the product in a specific way, i.e., to confer the 'effects' on a material or artefact, etc. Such 'effects', as distinct from properties, may be something as obvious as imparting to a material a resistance to combustion by naked flame, or to give another example; as obscure as a rapid cure rate in the case of a material such as thermoset composite.

This highlights an important distinction between categories of products, since many products are functional in their own right, i.e. they can be used to perform or fulfil specific requirements without further elaboration or incorporation with other products, e.g. materials, artefacts etc. Such products depend only on their own inherent properties to fulfil their requirements in service.

Some products are in a different category, i.e. they must be allied or combined with other products in order to enable their properties to achieve specific 'effects'.

An example of a product which depends only on its own 'properties' in order to meet an application requirement is one which fulfils the duty or function that the application prescribes for it without the need for further processing, addition or combination with other products.

Conversely, an example of a product which is dependent on being combined with other products to achieve 'effects' is a component of a system or system formulation, i.e. the 'system' performance is determined by the 'effect' conferred on it by the product.

It follows that product evaluation and definition must include both 'properties' and 'effects' in order to arrive at the most cost/effective product and its definition. At the feasibility phase, all of the relevant 'effects' may not be observed since it is essentially in the nature of the applications phase that subsequent 'effects' may be discovered by applying the product to those uses for which it may never have been initially envisaged.

The importance of distinguishing between products based on 'properties' and those based on 'effects' is reflected in Sections 5.2 and 5.3, where detailed examples in the form of case histories are provided.

We are now at the stage when a project needs its 'champions' to counter the well-intentioned attitudes of the business managers and marketeers who attempt to rapidly move a project from its research and innovative phases into the format of a commercial product. This inevitably means that such a project will receive an undue unit cost assessment emphasis instead of a cost/effective scrutiny related to application, i.e. before the project is mature enough to identify its real innovative potential.

When the innovator stands alone against the more conservative forces of the marketing organization he is particularly vulnerable, it is a price the innovator must pay but it does not always serve to advance his career. The choice is, do we keep our integrity and argue our corner for continued investment in a research project at the assessment stage or do we meekly bow to the business mind? I was in no doubt that the integrity of the innovator served his employer best when he argued his case for a project or exploitation route in face of opposing views. *In the limit the search is for truth in concept, experiment and business strategy.*

The latter remarks are not offered lightly, since in my own career I found that holding out for ideas that I was seeking to turn into innovative ventures meant that I may have missed out on some career opportunities, for as a senior manager commented some years later, something to the effect that, 'I knew that doing things in my own way did not necessarily ensure that my career interests were best served'. It was still in my view the only way in which I could work as an innovator and underscores that difference between horizontal and vertical thinking that to some extent separates the 'creative individual' from his colleagues and makes him incomprehensible at times to his own line management!

Throughout the assessment of the product we have seen the necessity for considering physical properties, application effects, manufacturing parameters' marketing influences and business management views with an emphasis on costs, but the key point that emerges is that the product definition has to be made cautiously in order to ensure that the merits of the 'new product' are clearly established and properly evaluated against the correct economic criteria.

As a project develops it becomes increasingly necessary to sharpen the criteria on which the case for continuing with the project must be judged, but a progressive build-up of information and confidence in the data will be obtained as the innovation advances to such decision points.

The evaluation of the process used to produce the product must be pursued in parallel with the product evaluation as the next section discusses. Both of these activities are indicated as Event 4 on the innovation project model, feasibility phase (Fig. 4).

4.6 PROCESS EVALUATION

In the experimental stage, much improvisation may have occurred in order to arrive at a rapid evaluation of the initial idea which generated the project. The priority at the feasibility phase is after all to demonstrate the feasibility of making a new product on which to base a new business venture, but as soon as the concept of the new product has been established it becomes imperative, in the case of manufacturing industry, to ensure in principle that the process used at the laboratory stage is capable of being scaled up to permit commercial production of the product.

The laboratory process can be broken down for consideration into clearly definable unit operations. This is immediately comprehensible to process design engineers who can then identify if the laboratory process can be readily translated by scale-up design procedures to match unit operation equipment which is already in commercial use and manufacture. If this search is positive the potential for manufacture at a commercial scale can be endorsed in principle.

Conversely, where the search to assess the availability of technology for commercial scale production which models the laboratory scale process unit operations proves unsuccessful, the option open to the innovator is to reverse the priorities by attempting to develop at the laboratory scale an alternative method of process manufacture, which does use wherever possible unit operations for which large scale technology does exist and coincidentally innovate new unit operations where needed to realize a novel new laboratory process which is now capable of scale-up with existing technology.

In some circumstances the new outstanding operations developed at the laboratory scale may still prove unadaptable to scale-up by existing technology, in which case a combination of innovation must be pursued in hand with process design that will yield in a novel new production plant concept, backed by new technology proven in the laboratory.

In both options, the opportunity exists for securing novelty of manufacture, either as a novel adaption of existing technology as in the first option quoted, or as the innovation of an entirely new process which uniquely provides the means and know-how for the manufacture of the product.

The advantage we secure for the project if process innovation is demanded in order to achieve commercial scale production is that process invention, in part or whole may be established. If we predict that this may lead to patents being established for the process envisaged it then reinforces the project's chances of successful exploitation especially if the product itself also may be patentable.

Of course at this feasibility stage the predictions we make will lack firm data on which to determine the probability of realizing them, but sufficient information will be established to indicate if scale-up is feasible and what degree of process design problem faces us.

As a consequence of the probable shortage of firm data on which to define the scaled-up process with confidence at this early stage, it will be inevitable that on attempting to evaluate product cost some approximation to order-of-cost tolerances will have to be made since the process economics determine the product cost.

A useful first approximation, which will be adequate for the purpose of making a broad initial assessment of the project, can be arrived at by simply considering materials and/or component costs which have been used in the laboratory scale production of the 'product'. In the case of a 'materials' type of product, multiplying this 'cost' by a 'cost factor', will cater for the associated 'capital' and 'operating' costs if other more substantive data is unavailable.

This kind of approximation immediately shows us if the product is likely to merit further consideration when compared with existing competitive product costs. We must note that our access to the latter is usually by first determining their 'price' and then we have to guess what profit margin the competitor is operating at in order to provide our rough first cost-assessment. We can extend our analysis of the competitive product costs by assessing its 'materials' and/or component cost and so determine what his capital plus operating cost factor is. In the 'materials' type of product I spent much time with, the cost factor was of the order of three.

4.7 RESEARCH AND MARKETING ORGANIZATION INTERFACE

In the early stages of a project, i.e. the feasibility phase, it is generally unwise to initiate any market research that is likely to alert potential collaborators, customers or competitors to the direction in which the project is developing, but some paper study can be of considerable value in establishing the kind and size of market to which the project can address itself. It is, however, essential to up-date marketing colleagues on the way the project is going, so that they can begin to get a feel for its scope for exploitation through discrete observation of the market place in general.

In the corporate research organization, this latter step represents an important scaling-up of interest and awareness in the project, while for the lone entrepreneur it raises considerable problems of resources in terms of gaining market data relevant to the project concerned.

At this stage in particular, it is essential to win the confidence and enthusiasm of the marketeers in the corporate organization, since without their co-operation the project will suffer from problems of liaison which can lead to break-down of communication with potential customers in the critical applications phase.

In preparing for the project review, when the marketeers will be brought fully up to date with the project status as it draws the feasibility phase to a conclusion, the events of evaluation, patenting, notional costing and market research on the basic product are of prime importance, but during the transition from concept to feasibility, ideas will have begun to emerge for optional applications for the product in its initial form and in secondary forms.

The marketing team in the larger industrial company can afford to allocate specialist marketeers to support new venture activities and divorce them from the existing business. Inevitably their target and priority is to see the new product in production as soon as possible and established on the commercial product portfolio.

This strategy is not without its drawbacks, since as mentioned earlier, a conflict of interest can arise between the marketeers and the innovators or researchers whose primary aim is to realize the maximum technological or scientific potential for the product, in the belief that the prime property of the product may not have been identified in the early stages of the innovation process, but may arise as the 'uses' are turned into new applications.

The consequence of the latter philosophy is for the innovators to attempt to keep a new venture in a fluid state while the applications are pursued, whereas the marketeers seek to freeze the project in order to align it with the necessities of commercial targets.

Applications often require a change in product specification in order to meet the 'use criteria'. This may not be initially clear to the potential customer or to the research and marketing teams, but once the innovator addresses himself to an application, the interchange of information on it between the marketeers, the potential user or customer and the innovator can precipitate the innovative process for secondary products; i.e. those products which are downstream modifications of the base product.

The innovation chain for secondary products is generally similar to that of the base product as we have seen in Section 4.2. As a result, the interactivity between idea, experiment and invention will follow the same pattern.

Clearly during the feasibility phase, the project will not be sufficiently advanced to cope with any extensive applications or market study, but it is obvious that any input of information made to the feasibility project review from these areas will increase the chances of the project moving into the 'applications phase'.

One factor that must not be overlooked during the laboratory stage of a project concerns the question of quality control. Much of the work on 'feasibility' is primarily concerned with the search for concept, properties or effects, but careful record of each experiment is essential if any results are to be replicated and confirmed. Such measurements eventually form the basis for quality control of the full-scale commercial product. This is a matter for increasing collaboration between the innovators and the marketing personnel, in particular when entering the applications and development phases.

In the case of the alumina fibre project referred to earlier, the key parameters for the quality control specification might have been seen as the thermal conductivity of the fibre mat or blanket, but as the next chapter shows, subsequent applications identified quite different parameters as the basis for quality control for the product and provide a clear example of the role and significance of the innovator/marketeer interactions as applications are advanced.

Quality control is discussed in more detail in Section 5.5.

4.8 PREPARING FOR PATENT APPLICATION

One question that occupies the minds of those who must take the decision to continue with the research and development of a new venture, emerges at the closing stages of the feasibility phase is: 'has the project any potential patent property?'. By implication this raises questions relating to product concept, processing or manufacture and applications, in which evidence of novelty is of prime importance.

The analysis of idea, invention and experiment as set out in Section 4.1 will, I hope, be found useful in determining clearly the nature of the innovative project under consideration and enable the innovator and patent agent to determine more readily if an inventive element exists in the project and if so to determine its characteristics. Recognizing what has been invented is not always obvious and any guide to achieving it merits consideration.

Should it be concluded that the project has identifiable novelty in its process, product or effect, it does not always follow that a patent application should be made immediately, since to use the colloquial terms of the patent agent, filing an application is to 'start the patent system clock ticking'; i.e. once the patent application is filed, the time allowed before its contents must be revealed to the public will start to be expended.

The importance of publication of a patent application is that it alerts the opposition or competitors of the direction and to some extent the progress made towards a specific product, which may constitute a threat to some existing enterprise by others. It also provides information on how the new product is realized and details the chief levels of attainment, in terms of properties and effects. This enables the competitors to start work on ways of circumventing the novel features of the new product so that its entry into the market can be challenged.

With the foregoing points in mind, assuming that no application has been filed, the subject of which is dealt with more fully in Chapter 7, it is essential for the innovator to attempt to formulate the required data and define his claim to novelty as soon as possible but without losing sight of the fact that a second stage of his work may eventually yield a more significant invention.

Compiling the patent claim and the data to support it sharpens the mind on what has or has not been achieved in generating the project in anticipation of the feasibility review and elevates the importance of the patenting preparation stage to one of considerable significance in the eyes of the innovator or inventor, as it can make a difference to the outcome of the appraisal of the project. The evidence of novelty is a major stimulus to both the research team and the business team.

4.9 COST/EFFECT CONSIDERATIONS

Providing the early cost estimates offer the possibility that the product will be cost competitive in the market and that the novelty of the product indicates that a patent is likely to be secured, it might be thought that a decision on proceeding with the venture is obvious, but other factors have to be given weight in the assessment procedure; e.g., does the project look as if it will produce a business which will fit the existing business and skills of the company? Some of these questions can be deferred to the applications project review at the end of the next phase of the innovation process as discussed in Chapter 5, but cost will be of increasing importance in the development of the project and it is appropriate at this point to emphasize the role of cost-effectiveness in assessing the merit of a new product.

The Alumina Fibres Project, referred to earlier, is a case in point. The unit cost of the fibre was greater than that of its competitors, but because of the absence of 'shot' in the fibre mat its thermal insulation properties were superior to those of competitors. In other words the thickness of insulation required for a given application using the alumina fibre will be less than that required for competitive materials and so the higher unit cost material can prove to be more cost effective.

4.10 PROJECT REVIEW OF THE FEASIBILITY PHASE

We conclude the feasibility phase with an important project review where all of the relevant information is presented. This usually takes the form of individual reports on each of the key aspects we have discussed in this chapter.

Typically such reviews consist of a project team meeting at which, researchers, marketeers, engineers, research economists and project manager are present. This is where the detailed reports are discussed and a decision is made on the project's future by the project manager who has financial responsibility for the research project as a whole.

The problems for the inventor-entrepreneur are of course very different at this stage.

The chief difference is one of resources; i.e., whereas the research team is backed by both financial resource in the form of research budgets and sanctions, as well as by a pool of skills to advance the project from feasibility to a prototype product via applications research and development, the lone inventor has none of these. The result is that it is vital to 'home in' on an application for the new product for which preliminary assessment suggests that a cost-effective and novel product appears to possess a strong prospect for success. This narrows the field of endeavour to manageable proportions and the lone inventor may be able to cope with the work involved in building up the various types of relevant data himself or be able to accommodate the cost of subcontracting it to others with a view to ultimately commissioning selected tests to be carried out by authorized test houses.

Typically in the UK we find that fire testing of materials for use in public buildings is undertaken by companies set up expressly for this purpose. Such companies possess the test rigs and measuring equipment to enable them to meet the precise requirements of UK and European Governments' regulations. Their results give credibility to the product and serve to open up a basis for a dialogue with potential collaborators who may respond by electing to support further research work or take up patent option rights as a basis for their own development programmes aimed at realizing their own venture.

In the absence of the support of a research team, the lone inventor needs to utilize the consultancy value of those who are ready to offer their views, but it is clearly a difficult road to travel, bearing in mind that the inventor must retain the confidentiality of his project.

One 'confidant' the inventor can rely on is the Patent Agent whose own code binds him to confidentiality on the subject of the invention he is contracted to file with the Patent Offices. The working relationship required between inventor and agent consequently provides an ideal environment in which to seek opinions on a project's status and exploitation strategy. In my own experience I found that I received much encouragement and constructive criticism on these vital areas of concern.

In concluding this chapter I would urge any lone inventor to secure for himself the services of a Patent Agent who will not only prepare a patent application from information provided by the inventor, but will carry out the implementation of the application and support him in some of the problems for which he finds the need to obtain professional advice. Not least in importance is the participation of the Patent Agent in any dialogue with potential collaborators or clients concerned with establishing Option Rights Agreements and other forms of contracts.

The project review for the research team is seen to be a highly organized activity but the review for the lone inventor in effect consists of two stages. The first is a dialogue between the inventor and his Patent Agent plus any private consultations he seeks under confidentiality. This prepares the inventor and his agent for a review under confidentiality with the collaborator with whom some level of discussions will have already been opened.

For the lone inventor the preparation of the exploitation proposal constitutes the culmination in many cases of his major contribution to the project, but in some cases he will either go on to explore product applications himself in order to increase the value of his patent property, or increase its credibility as a business proposal for appropriation by others.

Exploitation of the initial invention is seen to present the research team and the lone inventor ultimately with the same kind of problem, as the feasibility phase reaches the critical decision point for further commitment to the venture. Both parties must convince those concerned with funding the next innovation phase that the product is an eminently feasible proposition on which to build a new venture project and business.

5 Applications Research

5.1 APPLICATIONS AND SECONDARY INNOVATION

When a project reaches the point where it has been demonstrated that a new product is a feasible proposition, i.e. at the conclusion of the feasibility phase, we are faced with an important change of emphasis in the innovation process, incorporating both consolidation of what has already been achieved in the feasibility phase and the introduction of secondary innovation associated with product applications.

This change forms the substance of the applications research phase for which three objectives are paramount. Of these, the priority is to consolidate the basic new product by verifying the reproducibility of properties, effects, processability, etc. The second is to facilitate the base product's capability for satisfying specific applications which have been identified as targets for it. The final objective is the innovation of secondary products, which in general incorporate the base product in some form in their design, in order that the requirements of a wider range of applications can be met with a view to maximizing product and exploitation options.

Maximizing potential demand for the product, implied in this philosophy, is essential if an economic production or manufacturing process is to be established and if cost/competitiveness for the resulting family of products is to be obtained.

The individual events which make up the application research phase have been detailed in Chapters 2 and 4 and amplified by Fig. 5. This phase has much in common with basic research as described by the feasibility phase, but the relationship is a little like that which exists between Pure Mathematics and Applied Mathematics, i.e. the former addresses fundamental relationships and concepts, while the latter concentrates on a more 'practical' level which takes the findings of the fundamental study and then applies them to specific applications.

On equating this analogy to the idea of applications research, we find we must also include the fact that the efficient utilization of fundamental information in the way described, requires a specialist understanding of the applications themselves, i.e. in the case of product innovation, technological knowledge of the 'application' field is essential in order that secondary invention or innovation can effectively exploit the basic or fundamental features of the initial base product through secondary products.

The innovator specializing in applications research clearly has much in common with those engaged in basic research, but to my mind he is dependent to some extent on the intellectual resources, knowledge and innovation capability of the base product innovator, in the same way that we find the application mathematician is dependent on the pure mathematician. Conversely, the base product innovator depends on the applications ideas and creative contributions of the applications specialist to innovate the means of adapting his base product to often diverse uses and so facilitate its ultimate exploitation as a new business venture. Obviously on many occasions the innovator of the base product and the applications product will be the one and same individual. A particular case in point is the inventor-entrepreneur who periodically finds himself handling both roles as a matter of necessity.

In the research project team, where the probability exists that the respective products will be the responsibility of different innovators, the importance of close co-operation between them, almost goes without saying. In my own experience I can only witness to remarkably good relationships in this sphere and suggest that the company I worked with for some time ought to take some credit for the environment which they created and which made its New Ventures Group such a pleasure to work in for so many of us.

Applications research does, however, differ in emphasis from the basic research, in that the former is generally concerned with the creation of products to meet specific application targets, whereas the basic research is more usually concerned with exploring a basic field of science in which at the early stages, only a broad objective is identified and within which only a notional idea of the nature or form of a new product as a target may be envisaged, in which no more than a tentative view of its applications potential is discernible.

As indicated earlier, most of the applications research is implemented in the applications phase as exemplified by the Innovation Project Model (Fig. 3. and Fig. 5), but it also extends into the development phase and even influences the content of the initial product evaluation in the feasibility phase.

Once a specific application for the new base product has been identified, market research also can be initiated, drawing on the data provided by the feasibility phase with a view to establishing some idea of the potential market outlet for the new product. Availability of this information makes it possible to estimate a rough order of cost for the product in relation to specific applications. It is often found that the availability of 'research samples' is a desirable adjunct when seeking to secure market collaboration, after which the commitment of management and/or potential investors to the project can be explored.

With reference to the 'mathematical' analogy given earlier, familiarity with all aspects of the feasibility phase is essential for the innovator embarking on the applications research phase, as it represents the source of the 'production method' as well as the 'product data', which characterizes the base product. Both 'method' and 'data' are of fundamental importance to applications and their exploitation. Data typically consists of the key 'properties' of the product and quantification of any attainable 'effects' which might frame possible areas for applications. 'Production method' at the laboratory scale may indicate to the innovator that eventual scale-up for commercial production is also feasible and this will influence the development of application ideas and application product forms, or indicate the need for further process research. In particular the status of the basic process will have considerable impact on the way in which the production of a secondary or applications product will be contemplated.

Involvement of the Applications Specialist at the basic research phase is an important element in relating 'data' to applications. On occasions the Applications Specialist may recognize that a particular product might possess some important properties which have not been anticipated for it. Such properties can then trigger ideas for applications which open up a new field of potential use for the product. I recall one case in point, i.e. the resilience of inorganic fibre mats and their refractory properties. An appreciation of the potential usefulness of these properties when combined in a single product of this type suggested that applications subject to high temperature compression cycling could be secured for the particular product in question, as opposed to static refractory applications. Such observations turn our attention to aspects of product quality control, since reproducibility of 'properties' is essential for all applications.

Applications research is concerned with ensuring as early as possible in the life of a project that quality control criteria for the base product are firmly established. Clearly any quality control specification can only be tentative when initially based on the experimental data derived from the feasibility phase, which must be updated as the project matures. A start has to be made by assessing the tolerances attainable on critical parameters which characterize the products when produced by the laboratory process for sample manufacture. In addition, a parallel assessment must be made of the potential user's application tolerance requirements through market surveys and direct contact with potential customers.

Quality control specification compilation is a matter of reconciling the two aspects identified, i.e. production tolerance capability and user's tolerance requirements. The two sets of tolerances will include common criteria, but also factors which are exclusive to achieving specific 'properties' at the production stage, while others will relate to the method in which the product is used in a specific application in order to achieve a reproducible effect or performance in service conditions.

Also in association with the latter, the requirements of regulatory bodies who set quality and performance standards must be complied with, e.g. British Standards Association, international and national regulatory bodies of many kinds, Trade Associations etc., depending on the field of application and market identified for the base product and ultimately, the secondary products.

A deeper consideration of the various aspects of quality control are provided in Section 5.6, but it is introduced here to show how efforts to establish it precipitate the emergence of secondary products, i.e. it is precisely because the initial product cannot necessarily meet the quality control specification requirements of every application it encounters, that secondary innovation takes off within the project.

Of course secondary innovation and the resulting applications products arise through other circumstances also. In some cases the concept of an alternative form of the initial base product leads to a new application, while in others it is an idea for an application that precipitates the creation of a secondary product to meet the identified need, but as we have seen from Chapters 2 and 4, the relationship between idea, invention and experiment, as the initiators of the innovative process means that such a definition of the origin of applications products is an oversimplification. In fact their origin is open to the same degree of options as indicated for the base product.

Where a secondary product emerges that does not incorporate the base product in any way, the result is a new basic product and by inference a 'new process'. We then find that we have started a new innovative chain in which both product and process must be exploited as a new and distinctive business venture. *This in effect represents an innovation 'shift' which results in either the abandonment of the initial base product innovation or the pursuance of a parallel evolution of it alongside the secondary new basic product,* as Fig. 11 shows.

An important aspect of those applications that give rise to secondary products is that while providing a means for successfully exploiting the initial or base product they also enhance the potential growth position for the business venture as a whole, due to the inherent flexibility in exploitation strategy provided by a product that has a number of different uses and consequently markets, on which to secure its sales portfolio. The more diverse the range of applications which can be identified at this stage, the greater the confidence in the potential of the project. In many cases it is the applications research and the feedback from the market that gives credibility to the venture in the eyes of the investor or company management during this phase, but excessive diversification can be distracting as discussed later.

To highlight the difference between the base product and secondary products; by way of illustration, a domestic microwave oven is a base product whose use is predefined by virtue of the design parameters built into its concept and specification, i.e. its application is inherently defined and in general it is not likely to form a subject for consideration in secondary applications unless they have been anticipated in its 'design led' innovation.

The illustration given above also emphasizes the distinction between 'design led' innovation, generally targeted on a specific application, and that of a conceptual or experimental led innovation, often including invention, which is not initially targeted at any precise application but on the realization of useful 'properties' and 'effects', from which a generic group of applications and products can be expected to arise as discussed in Section 4.5.

This latter point brings us to the importance of determining early in the life of any new venture if the dominant factors which characterize the product are its 'properties', or its facility for creating 'effects'. Sections 5.2 and 5.3 provide illustrations of both types of project.

5.2 'PROPERTIES' AS PRODUCT BASIS

The objective here is to demonstrate from project case histories, the distinctive nature of the role of 'properties' in characterizing a product. If we take for example a material that possesses low thermal conductivity, and structural integrity typified by mechanical strength data and then compare it with the properties found in other materials that also possess low thermal conductivity, we are immediately presented with an indication of its application potential as a competitor in, e.g. the building insulation market where the existing materials may already find their application.

With this information, we might characterize the material as 'a rigid thermal insulation' potentially suitable for manufacturing in slab or board forms, i.e., we will have characterized the product from its 'properties' and with reference to its potential market. The product therefore offers immediate and direct application targets without depending on any other process operation to introduce usefulness, but of course many other properties must be evaluated to see if the 'new product' exhibits technical advantages or shortcomings with respect to the competitive materials.

It is evident from our initial assessment of the 'new product' that any inadequacies we discover in it will become the focus of efforts for improvements, which form an important part of the innovation chain. As Gardiner and Rothwell (1985) say, 'innovation is an iterative and incremental process involving continuous redesign and re-innovation'.

A practical example of the above, is drawn from a period when I was engaged in supporting a project team whose objective was to develop, produce and exploit a new refractory fibre product i.e. an alumina fibre, referred to earlier. This project typifies experimental led innovation, in which the product's 'properties' define product functionality to a large extent.

In describing the early stages of this project I have concentrated on the interaction of the base product and secondary products to illustrate that while 'properties' define the base product, they are often the stimulus for the inception of secondary applications and the characterization of secondary products.

The basic and application research for the 'fibre' is detailed by Birchall *et al.* 1985, in the *Handbook of Composites* (Watts and Perov, 1985). The project's initiation arose from an appreciation of the potential of inorganic materials as precursors for making strong polycrystalline refractory fibres with high temperature properties.

Innovation was initiated through experimental studies of fibre processing using a variety of feedstock materials. The feasibility of a particular formulation and method of processing alumina fibre with a high silica content was demonstrated and its basic properties were determined. This provided the base product from which the applications research was to develop.

In case this narrative of the inception of innovation of a major product sounds somewhat bland I must stress that it was a stimulating experience for many of those working close to its source and for others who were out on the fringe of the endeavour, perhaps only able to contribute in a relatively remote way to the project. It was such an environment that reached me while I was locked into a career concerned more with managing the services for innovation than participating in it.

To the creatively inclined, in the latter situation such an encounter with an innovative new venture can provide the spur to switch the emphasis of one's own career in order to fulfil the innovative drive also. In my own case making such a change proved to be difficult. Any attempt to change direction seemed to be deliberately frustrated by a management that was not fully alert to the potential role of engineers as equal partners with scientists in the field of innovation, but sometimes events precipitate an opportunity for change which are quite unpremeditated and when they do, it is time to take the initiative as later comments indicate.

From those early experiences we all learned to some extent to respect each other and appreciate that we each have our limitations as well as our strengths. It was a highly formative period for the New Venture Group as well as for its individuals which emphasizes to me that establishing a New Venture team is an innovation process in itself! In retrospect it is difficult to describe the sense of excitement and satisfaction that overtakes the innovator when at last he finds an environment which actively encourages his attempts at making a creative contribution towards a new venture, typified by the way that the impromptu pattern of the day led to a continuous interchange of ideas, problems and debates on innovative avenues of progress to exploit the newly conceived projects.

I had the good fortune to be surrounded by lively knowledgeable and good-natured colleagues with whom it was a pleasure to work! In particular the project team gradually drew together a sprinkling of creative people who were complemented by the analytical capabilities of their colleagues. The innovative environment was eventually excellent and this was in no small degree due to one or two managers who opened the door of a research department, highly preoccupied with product improvement, to the alternative opportunities of new ventures and new products. My much more illustrious namesake Dr F.R. Bradbury was instrumental in redirecting much of the thinking in the Research Department to the new philosophy for research which stressed the benefits of high added value low tonnage products as a basis for business exploitation,

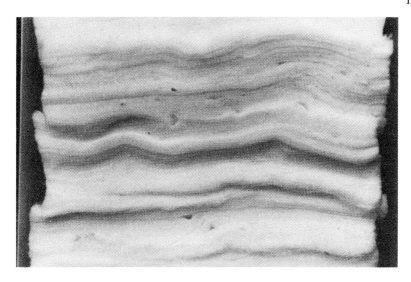

Figure 13 Inorganic short-staple fibre mat (Reproduced by permission of ICI Plc)

instead of the preoccupation of the day with high tonnage process/cost improvement research, which at the time was yielding low returns on research revenue and capital investment.

In the case of the Fibres Project, which was at the centre of these changes in research emphasis, subsequent work led to the decision to develop the product as a 'short staple' fibre instead of in the form of 'continuous filament' fibre. This entailed a collection process for the spinning arrays that resulted in a 'mat' of randomly distributed fibres as the final product form (Fig. 13). In this 'edge on' view the lamella construction can be clearly seen.

Characterization of the base product was consequently determined by the process of manufacture, i.e., the base product was defined as a 'random fibre mat', whose 'properties' were derived from its alumina silicate structure, physical characteristics and distribution in a specific bulk form. The product could then be seen to fall into a generic group of materials which immediately influenced ideas for its use, applications and exploitation, i.e. it was seen to fit in a group of existing *low density inorganic thermal insulation materials of mat or blanket form.*

These observations immediately provided the initial base product definition and also precipitated a search for applications in which the unique physical 'properties' of the fibre could be exploited, but as pointed out above, the generic group in which the product fitted exerted a powerful influence on the direction of this applications search; a field in which I was able to make a contribution as an engineer viewing the project from a different standpoint from that of many of my colleagues who were engaged primarily on the base product research.

The initial applications were naturally directed at the more immediate potential uses for the fibre in the form in which it came off the semi-technical process plant line; i.e. as a non-woven blanket or mat of fibres, which was similar in form and functionality to several other competitive short stable fibres available in the market place for prime use in thermal insulation, but the alumina fibres possessed some distinct advantages over competitors. In the initial stages of the project these advantages were thought to lie chiefly in its superior thermal stability, with the result that the primary application was almost self-determined as a thermal insulation blanket for high temperature applications in industrial furnaces, among which were large pottery kilns, etc.

The need for other applications became imperative in order to increase the potential market for the product and improve the economics of production. This entailed the research team and the marketing team in searching for further areas where high temperature fibres were already in use or being contemplated for use. It is this kind of input that makes it possible to assess a project's chances of success and determines the degree of continued research commitment to be made to it, at this early stage in the innovative process.

It is evident that a low density mat with high thermal stability should find outlets in high temperature thermal insulation, but because the fibre was of such high aspect ratio, which conferred on it distinctive flexibility, it was soon found that this latter feature could restrict its use in applications where dynamic gas conditions are met, primarily due to the risk of fibre erosion or loss of fibres from the mat. An attendant feature of the mat which also made it vulnerable to erosion was its lamellar construction, i.e. the fibres formed non-woven lamella on collection from the spinning arrays, but the interlamellar 'bonding' or adhesion was weak and as a result the top lamella could be stripped away in the severe dynamic conditions referred to.

The identification of such limitations and their cause, is of course of considerable importance. Such a situation immediately leads the innovator into searching for ways of overcoming erosion, or of developing a secondary form of the product which can withstand it. In this way it was hoped to secure a place in the market for the base product or a secondary product form in the applications which prove to be most appropriate to them.

Very briefly, one of the solutions to the erosion problem led to the obvious step of adding a 'binder' to secure the fibres together, which of course resulted in a product of higher density and entailed considerable end-processing to produce a new end form, i.e. a rigid thermal insulation board, similar in appearance to many other insulation boards. Inevitably this solution implied significant increases in the cost of production since board making involves the use of presses, moulds, and slurrying equipment, plus the added manpower needed for process operations.

Although this secondary product was not necessarily novel, it was an innovative step which catered for the limitations in the base product 'properties' and in so doing provided a wider range of applications and markets for the product. It is a clear example of the utilization of the base product to provide a secondary product by 'end-processing', but highlights the fact that the assumption that the initial base product will meet the requirements of the market in its initial form is often a complete misconception.

It is obvious from the above that considerable difficulty may be found in achieving an application for the base product which does not require some secondary innovation when encountering market conditions, since the initial application anticipated for it is often somewhat rigidly determined by apparently logical considerations, but with limited knowledge of practical application needs. A wider more transverse approach to applications is essential and this needs to be introduced at the later stages of the feasibility phase if possible, i.e. in anticipation of the impending needs of the applications phase itself.

The development of a fibre board to resolve the erosion problem in the particular case of the 'fibre' project serves to highlight the disadvantages that are attendant on some secondary products, such as higher processing costs and a reduction in cost/effectiveness. In this case the manufacture of a board carries with it the consequences of increased density, compared with those attainable in a 'mat' or 'blanket' form. This increase in density implies that the board thermal insulation efficiency is less than attainable with similar thicknesses of 'mat', i.e. the boards are less cost effective.

Fortunately another solution had already been identified by other workers in the refractory fibre field, in the form of a stack-bonded fibre panel or tile, which is shown incorporated in the construction of a pottery kiln ceiling in Fig. 14. This fully utilized the basic mat form and properties and did not entail the use of the more costly wet processing methods of board making in its production, i.e. the resolution of the problem was a 'macro design solution', instead of any fundamental change in the matrix of the fibre mat. Standard mat was simply cut into strips and stacked together, to form 'titles', by securing the stacked material with an organic scrim, in which form it could be applied to walls and ceilings. The scrim was burnt away in service without impairing the lining which had been created.

It is important to note that the innovative team had no prerogative in resolving applications problems! One obvious benefit of researching in a generic field.

The latter example illustrates that a new product can emerge which apparently falls short of requirements in some physical property, but which on further reflection is seen to be a limitation imposed not by 'matrix properties' but by 'structural properties', i.e. the inadequacy lay with the failure to integrate the fibres in such a way that they did not produce a lamella structure. Resolving this is easier said than done, but the alternative was to undertake applications research with a view to circumventing the inadequacy, not by re-inventing the product with different properties, or even by indulging in costly end-processing conversions of its form, but by using the product innovatively in its basic product form.

We have now introduced the role of design as a means of making the initial base product satisfy the application need. Obviously in this case the fact that

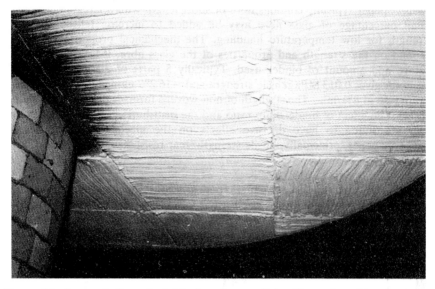

Figure 14 Inorganic fibre stack-block — furnace lining (Reproduced by permission of ICI Plc)

the mat was of lamella construction by virtue of the manufacturing process employed, rendered it vulnerable to delamination in turbulent gas conditions, or as we have described it, to fibre erosion. The stack-block construction turned the lamellae through 90 degrees and by some compaction of the mat in the forming of the block, which was retained on assembly, it ensured that the lamellae were adequately secured by the compression between the lamellar faces while at the same time the exposure of the lamellar faces was eliminated by the construction used. The exposed face consisted of fibres which were interwoven in the plane normal to the exposed surface and as a result could withstand the gas turbulence.

The example points to a key decision point in the innovative chain, that is that the initial identification of the major properties possessed by the base product clearly requires careful consideration since in the case of the fibre project, the more obvious property of high thermal resistance might have proved to be less commercially and technically significant than, for example, the mechanical properties of the mat, i.e., its fatigue and resilience characteristics or perhaps its tensile strength.

It is obvious from these examples that the initial identification of the 'properties' provided by the product, determines the basis of the economics of the project and leads to targets for the applications and development phases. It also illustrates the inception of secondary products in order to exploit the base product 'properties' as an advantageous way to achieve product exploitation through secondary end forms. Or put another way, we have seen that in some situations, *design and not 'properties' provides the key to the wider application of the base product and generation of the secondary products.*

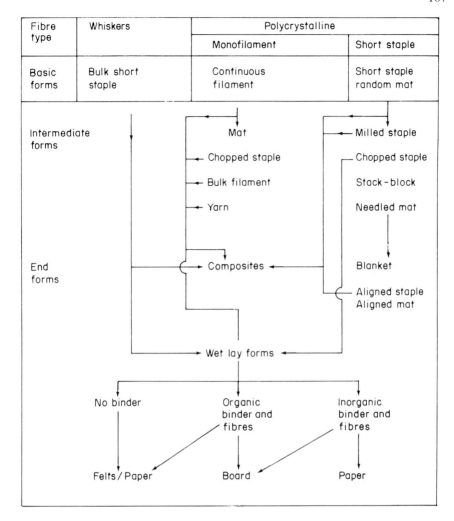

Figure 15 Alumina fibre — from base product to end forms (Reproduced by permission of ICI Plc)

We find an example of the relationship between 'properties' and design in the case of the innovation of products such as inorganic fibres described in the *Handbook of Composites* (Vol. 1 Chapter IV, by Birchall *et al.*, 1985), where the innovation of end forms based on both polycrystalline fibres and whiskers is illustrated by Fig. 15. In this case intermediate forms are first of all realized by processes that do not involve any basic change in the fundamental structure of the product, only a reorientation of the fibres in different geometries. However, in the case of the end forms, design plays a part in the introduction of secondary processing and in some cases the introduction of additives, e.g. binders, to produce a secondary product with fundamentally different properties from those exhibited by the basic fibres.

Some of the resulting end forms have already been referred to and others arise which in the specific case of the alumina fibre project, involved a further design contribution in order to relate the new end forms to specific engineering applications.

5.3 'EFFECTS' AS PRODUCT BASIS

The alternative type of new venture project we encounter is where 'effects' constitute the exploitation feature, i.e. where the 'effects' attainable by the use of the product in a specific application or formulation, determine product value. These 'effects' of course, derive from the properties, but once we have identified, preferably at an early stage in the innovation processes, that 'effects' are likely to determine the product exploitation strategy in any business venture with it, *the essential feature of further research will be to correlate properties with effects.*

This change of emphasis requires a dual and parallel test and evaluation programme. Products based on 'properties' such as the fibre referred to in the previous section are evaluated by determining inherent behaviour characteristics against loading conditions. The limiting values of the 'properties', e.g. strength etc., determine the usefulness of the sample, against which a whole range of applications can be contemplated. By contrast, although products based on 'effects' must still be subject to the measurements of 'properties' which are inherent to them, in order to eventually exercise quality control at manufacture, they also require an evaluation of behaviour characteristics when placed in an environment which now is concerned with the synergistic behaviour of the product in relation to the materials or components with which it forms a part. In other words the 'effects' product can only be fully evaluated by measuring how the *system* in which it is incorporated responds to service conditions as presented in a specific application.

The example I have selected to illustrate the differences this makes to the exploitation of 'effects' based products is drawn from the field of composite materials, in which the innovated product is identified as a potential constituent in the formulation of resins used in the manufacture of the composite. The merit of the product in this case is to be found in its deployment as part of a system on which it confers specific benefits.

Initiation of company interest in composites arose as a result of research organization and technical planning efforts to explore outlets for the excess production capacity of a particular chemical plant, by attempting to identify uses where it could serve as a precursor or feedstock for processes employed in the manufacture of other products. In this case it was recognized that the particular chemical concerned could be used in the manufacture of bisphenol S, a chemical that possessed similarities with bisphenol A, which is used in the manufacture of epoxy resins and a number of other products.

The marketing and technical departments established that such a product as bisphenol S should find outlets in areas currently met by the use of bisphenol A, providing cost advantages were attainable and technical performance was satisfactory.

Contemporary applications for bisphenol A were identified, but the one which was of greatest interest was in the manufacture of epoxy resins and in particular with the resins used in the production of carbon fibre epoxy composites. This application proved to be of particular importance to the aircraft and automobile industries where weight saving and high strength/mass ratios were desired. Evidently fuel economy, weight and performance were the focus of interest. This was all consistent with the observed market trend to displace metals with composite structures employing fibre reinforcement to achieve the economic strength/mass characteristics.

The target was consequently set of determining the prospects for displacing bisphenol A with bisphenol S in such epoxy formulations and as it later appeared, to extend the potential of the composite manufacture by *improving the economics of production*.

The role of the innovator in a project of this type, involves initiating research studies of the properties attainable from the use of the new product when utilized in the manufacture of composite resins. It has to be noted that it is not simply the resin which is of interest, but the 'composite', which consists of reinforcing fibres in a resin matrix. The innovator in this instance has to learn from and depend upon the expertise of specialists in the field and analyse the result of their tests on resin samples and ultimately on fibre composites themselves, i.e. the ultimate task is to assess the 'effectiveness' of the composites which utilize the 'effects' product in their formulation. To complete this picture we remind ourselves that the results must be correlated to the quality control of our own product at the manufacturing or laboratory stage.

The other significant action required is to monitor the cost implications of changing from one feedstock chemical to another. Clearly if a cost/effect assessment were to be made purely on a unit product mass/cost basis it might show that the new product was not sufficiently competitive to displace the existing bisphenol A product.

In the case in question, sponsored research by specialists in the field of composites revealed that bisphenol S based epoxy resin displayed comparable physical properties to the bisphenol A epoxy resin — but most importantly the bisphenol S epoxy could be formulated to 'cure more rapidly' than the bisphenol A resin while still retaining its other competitive physical properties. The significant point here is that the potential for 'high speed cure' is an 'effect' conferred on a system or process and constitutes the characterization of the product as an 'effects product'. This meant that the product's attraction for the market place lay in its potential for improving the economics of the industries composite process and its cost/effectiveness has to be read in those terms instead of as a product in isolation at the factory gate.

As we have seen, the latter point transformed the potential of the bisphenol

S product from a 'commodity chemical' into an 'effects chemical' and the exploitation strategy was then concentrated on offering the market a 'high speed cure resin formulation' for mass production processing of large reinforced composite components.

The point to be made is that an 'effect' which might be obscure to the casual observer, had the potential for use in a rapidly growing market, where added value could be recognized in the product, i.e. in the field of fibre reinforced composites, which are the basis of fabricating strong, tough, structural components for aircraft, automobiles, etc.

The fact that weight saving over metal components was already attainable by using bisphenol A based composites and that similar benefits would be expected with the use of the bisphenol S product, indicated that any reduction in processing time and cost associated with manufacturing the composites would be extremely beneficial and indicate that a significant product advantage might be secured. Hence the interest of the aircraft and automobile industries in the potential of the product under research, at the time.

Sheet moulding compound processes are used to form many of the fibre-reinforced structural components, but this necessitates the use of large presses which are very costly to build, install and operate. Maximizing the production capacity of such presses is evidently highly desirable in order to reduce costs. The introduction of a high speed cure is critical to reducing the residence time of the component in the press and this formed the basis for the approach to the market research.

One of the features of this project was that the research chemists recognized the dominance of engineering in any product research and this was also endorsed by the marketing team, with the result that the market research and applications research were considered as being so intimately interwoven as to require both functions to be carried out by an engineer reporting back to the technical management for the product area.

Before we dismiss this latter account, as of incidental interest only, let me point out to the uninitiated that for different professional disciplines to be able to work together in the way described is a triumph in any organization. Much credit goes to the manager who influenced the level of collaboration which was secured.

Returning to the technological aspects of the project, the anecdote serves to emphasize that we have to be constantly alert to the key properties of a product which inevitably will control its application potential. In the case quoted the mechanical strength and other physical characteristics and properties of composites utilizing bisphenol S showed little advantage over those of bisphenol A, but the value of high speed cure was a major potential selling point since it offered the possibility of major cost savings in production, not of the resin, but of the resulting composites!

The key property was thus only detectable by going 'downstream' in the exploitation chain, until the application emerged that could take advantage of an 'effect' which would easily have evaded inclusion in the compilation of the product data sheet, on which basis product assessment is largely made.

Sadly the project foundered when further cost studies revealed that the potential for cost competitiveness of the manufacture of the bisphenol S was not as strong as was first forecast by the designers of the process. This, however, serves to illustrate another feature of the exploitation and assessment stage, which is that economic judgements can so easily overlook the innovation consequences and vice versa.

In this case the fact that the bisphenol S was more expensive to produce than first estimated was not the only basis on which the project should have been assessed. The issue was; could a high speed cure epoxy resin 'composite' based on bisphenol S be produced more economically than could be achieved with the bisphenol A based composite? It is clear that 'effects products' must be assessed economically at the point where its 'effects' come into use. In this case it should have been at the composite manufacturing stage. Of course, developing a satisfactory exploitation strategy for the product becomes increasingly difficult the further downstream we have to go in securing its application and the exploitation of the product's potential for contributing 'effects'.

The example surely emphasizes the need for great care in the feasibility phase, especially during product definition and assessment, to avoid the possibility of decisions being made on the project's status without anticipating the next phase of innovation; i.e., the applications phase, in which further data may alert us to the wider potential of the product in terms of 'effects' and so prevent us from discarding the project prematurely, or erroneously defining the product in a way which will inhibit its subsequent effective exploitation. Such applications as the 'composite' serve to emphasize that *it is cost-effectiveness at the application end of the innovation chain that must ultimately determine the potential of a new venture of this type.* Even now I wonder if we rejected the project in error by failing to follow the kind of philosophy advocated above!

5.4 APPLICATIONS AND MARKET DEVELOPMENT

One way of looking at the function of applications research is to consider it as extending the innovative process into the market place in order to relate know-how of the base product to the needs of both application and market. In consequence, we find a critical interface within the innovation project model where the innovator, or more particularly the inventor, is involved with close liaison with the marketeer. This brings to light differences in philosophy which are inherent between them.

I believe these differences are to some extent caused by the traditional attitude of the marketing organizations of some large companies, which tend towards treating all contact with potential customers or other external bodies involved in applications research as their exclusive domain.

Clearly the intrusion of the inventor into market contact situations makes the marketeer's hair stand on end, since he is only too aware of the way in which inventors can react to market exploitation problems, by introducing solutions which change the course of the project and transfer the initiative from the marketeer back to himself.

The marketeer on the other hand, is always seeking to finalize a product definition and rationalize it into the form of an established commodity on the sales portfolio so that it can then receive all the benefits of established products, i.e. the technical service and promotional resources of the company.

As with most of us, the marketeer will also be motivated by self-interest, in particular where it appears to support his career aspirations, as can be the case when the establishment of a new product provides openings for his promotion into an influential position on sales or business management. It is obvious, that while the innovator seeks career fulfilment in keeping a project open to the opportunity for introducing further innovations, the marketeer seeks by contrast to minimize such situations in order to provide himself with an opportunity for obtaining greater autonomy over the project as early as possible, with a view to implementing his own ideas for business development.

Both of these interests have to be respected, but the prime aim of achieving the exploitation of the product must not be over-simplified, since successful exploitation will require continuous alertness from its business and research managements in order to keep both objectives alive. In fact the need for a degree of sustained tension is likely to ensure that a proper balance will be kept over the interests of further innovation and of consolidating the project in market specification terms.

Efforts to freeze the product specification at too early a stage in the development of the project runs the risk of introducing a product to the market with too little flexibility for responding to market needs. The consequence of which is to lose opportunities to secure an outlet for the product which might need to be gained by 'secondary innovation'. Experience shows that the initial idea is not necessarily the one which will characterize the outcome of the venture. It is often secondary ideas and inventions which do this precisely because the inventor has reacted to the needs of the market place with secondary solutions to problems of specification or application.

Achieving a satisfactory balance between the creative response of the inventor and the economic aims of the marketeer in product exploitation is clearly a matter for much concern and alert management. It could be argued that management's role is to ensure that an efficient teaming of inventor and marketeer is achieved by a consideration of the personalities involved, but the professional approach to the problem is to analyse the differences in the attitude of both disciplines to achieving the team's objectives and to show its members where misunderstanding can arise as a result of their differing philosophies. These differences can subsequently be focused upon at the periodic project review meetings should day-to-day management and communications fail to achieve the desired integration of effort.

I have found that while exercising a creative or transverse approach to problems and target realization it has become necessary to attempt to implement creative ideas and invention through the innovation process in a disciplined and logical way. Not all marketeers recognize the duality of this approach and the innovator's periodic obtuse departures from a logical development may seem incomprehensible to them. I can sympathize with their reaction. It is often incomprehensible to me also, especially when attempting a case study with a view to defining the source and route of invention in particular. I also believe that if both parties share an appreciation of the way in which a new venture project develops in the form of creative or transverse initiative interacting with logical progression and management, they will find that synergy between them leads to greater efficiency and harmony in new venture enterprises.

A fascinating illustration of the innovator/marketeer interaction was afforded for me when attempting to find an outlet for 'Saffil' alumina fibre in the automobile industry, as the following reveals.

The Fibres Project had identified that the USA Environmental Protection Agency had at that time imposed severe controls on the level of automobile gas emissions from exhaust systems. As a result, catalytic convertors were to be developed to fit in the exhaust systems to reduce the level of nitrous oxides in particular, in order to meet the need for reducing the toxicity of emissions.

Because of the high gas temperatures involved it was necessary to distribute the catalyst, which was to effect the reduction of the oxides, on a porous refractory form, referred to as a 'monolith'. Some extruded alumina-silicate refractory monoliths were already being produced by a number of manufacturers, but it was thought that a competitive product could be realized by utilizing a 'secondary product form' made from 'Saffil' fibres, i.e. a 'Saffil' paper, incorporating selected binders to produce the monolith. This was expected to provide a high specific surface area which would ensure superior exhaust gas/catalyst contact and also be capable of withstanding thermal shock, known to be a problem affecting the competitive extruded monoliths at the time.

The research team produced both 'paper' and techniques for forming the porous monolith. It was a brave effort to achieve an advanced end form, involving not only the development of the alumina fibre paper, but the design and construction of a monolith which required both flat and corrugated paper to form the exhaust gas channels in it.

All of this task had to be carried out against entrenched competition and a deadline, set by the EPA, for introducing the monoliths into service. In addition to these problems the market interest was dominated by the competitive products which had already established an entry into the long process of 'product certification' at a much earlier date. These alumina silicate monoliths possessed particular production advantages over the 'Saffil' monolith, which meant that the legislative target 'compliance date' set by the EPA could be met by them.

The combination of the various advantages offered by the competition, the time scale and production demands of meeting the EPA compliance dates for certification and the late entry into the 'market race' of the 'Saffil' monolith

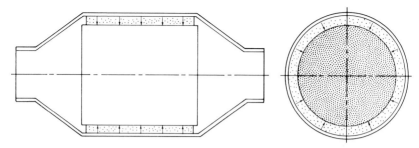

Figure 16 Monolith fibre support — cross-section (Reproduced by permission of ICI Plc)

eventually led to its abandonment. This represented a significant decrease in the potential market for the base product in a 'secondary form' i.e. as 'paper' and threatened the prospects for the Fibre Project itself.

The innovative response we then made was that the same tonnage requirement for fibre in the catalytic converter project could be found in deploying the fibre in a different role, but still as part of the converter construction. This was to use the 'Saffil' fibre mat as produced by the base product production process to provide a 'monolith wrap' which would support the competitive alumina silicate monoliths and enable them to withstand the dynamic and thermal conditions which they had to endure in the exhaust system.

What was surprising was that while all eyes were on the use of the fibre in the monolith, the concept of supporting the monolith in its high temperature environment had not been contemplated as a market opportunity. In fact the mass of the fibre required for supporting the monolith was significantly greater than that needed for monolith fabrication!

The new application was sanctioned for exploitation. It had an advantage over a number of alternative applications in that the contemplated product form was closely allied to the basic product which was the 'mat' form. This meant that the extent of additional processing was confined largely to quality control of the base product and the ultimate development of a containing envelope which would ensure that each 'wrap' consisted of a specific mass of fibre.

Innovation through design was to play an important part in the application of the new end form, which had to meet a rigorous specification which included providing a gas barrier seal, as well as being resistant to fatigue loading conditions when supporting the refractory monolith. The features of the design were presented at a symposium on automobile automation (Bradbury, 1977).

The insulating seal was to be made between the metal shell of the converter and the refractory monolith carrying the catalyst. To withstand the longitudinal dynamic loads on the monolith arising from the engine and transmission vibrations and inertial conditions, woven metal wire and rings were included in the initial design, which in every other respect is as shown in Fig. 16. However, early tests of this 'fibre and woven wire system' in our own laboratories and with the automobile companies revealed repeated failures in the form of fibre loss and monolith collapse. This initial pattern of failure is all too typical of many new products, but the reader will recall that redesign and re-innovation is the order of the day in new ventures! In this case it was concluded that the rings were inappropriate as they contributed to the break-up of the ceramic monolith.

Business opportunities for the 'Saffil' fibre were now threatened and a new response was required. To have urged the automobile companies to change their converter design while freezing the fibre specification would have rapidly concluded our efforts at exploitation of the product. The alternative was to offer a new way of utilizing the fibre in the assembly by dispensing with the wire rings and maximizing the retention capacity of the fibre wrap by increasing the compressive radial forces between shell and monolith.

We conclude that by *increasing the packing density of the fibre in a controlled way to provide both insulation and mechanical restraint for the monoliths*, sufficient restraining force would be developed to withstand the dynamic loads on the system without the necessity of using the wire end rings. The final design was consequently produced exactly in accordance with Fig. 16, i.e. omitting the end rings.

At this point the marketing team's reaction might have logically been to withdraw from the exploitation of the product for this particular application, but happily the co-operation in this instance between the team and the marketeer was very close, the marketeer displaying great initiative and not a little faith in the inventor, by backing up the proposed innovative response to the problem encountered.

The marketeer's role was also critical in securing the continued collaboration of the design team of the potential customer on the evaluation programme with a new design concept which recognized that the fibre had to be utilized in a special way for it to perform its task and produce the effects expected of it.

This kind of innovative response demands some understanding of the properties of the base product and its potential in the application proposed. It demands careful judgement by both innovator and marketeer in deciding to risk an untested product in a test and evaluation situation controlled and observed at first hand by the potential customer! The opportunity had to be seized without delay or be lost immediately to the competitors.

Clearly it is wise to devise some simple type of experiment or test to be carried out under one's own control and on home premises in order to screen the product to be offered to the 'customer'. In this way the commercial as well as the technical risk of catastrophic failures are minimized and the chances of keeping the goodwill and confidence of the 'customer' is increased. In this case the test devised was to assess the compressive and fatigue properties of the fibre in a form which modelled the exhaust system configuration and to some degree its loading conditions.

This example is typical of 'secondary innovation'. It entailed presenting the fibre in the form of a carefully calculated mass made up in a convenient pad form for assembly purposes. The mass determined the final packing density on assembly, in relation to the manufacturing tolerances of the shell which the automobile company had already defined. The result was a system which could achieve the restraint of the monolith due to a quality controlled assembly method that ensured uniform and consistent retention forces on the monolith in service. Section 5.5 expands this in more detail.

You may well ask why all this fuss over the packing density. The reason was that the pressure exerted on the monolith was directly proportional to the final packing density. The product in the form of the 'monolith wrap' had to accommodate not only dynamic loading conditions, but wide variations of temperature as well. Only by satisfying these conditions could the system stand any chance of achieving a market breakthrough.

Since the relationship between the fibre properties and the design of the support system was critical it was essential to carry out carefully designed test procedures in order to ascertain the properties at various loading conditions. This process of 'testing' lay at the heart of project.

In the case in question the key parameters of density, fatigue stress and thermal stress degradation were all evaluated with the co-operation of the Ceramics Research Association and with UMIST. The results are shown in Figs. 17, 18, and 19.

In the event, the new system was effective but at a later phase failed to be adopted by the potential customer possibly due in one particular case to subsequent problems of collaboration when the management of communications changed hands and vital manufacturing criteria were not successfully conveyed to the evaluators, but that is another story.

To show how difficult it is to secure the final adoption of a new product in these situations, one of the potential customers for the product accepted that all the technological performance requirements for their converter design was

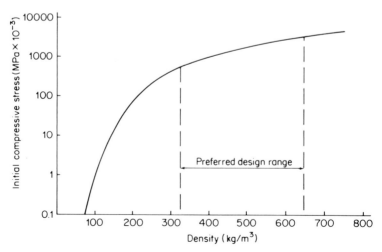

Figure 17 Monolith fibre support — comprehensive stress/density (Reproduced by permission of ICI Plc)

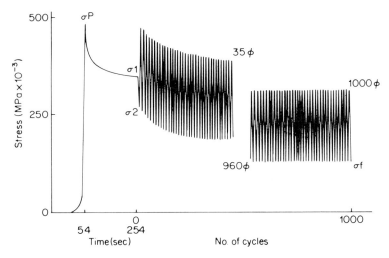

Figure 18 Monolith fibre support — fatigue stress/cycles (Reproduced by permission of ICI Plc)

met by the data we presented to them, but the failures they were experiencing with a competitive system were not sufficiently great in number to justify making the change to our superior design, especially as they would face an extensive series of re-certification testing to comply with the EPA of the United States in whose market they were operating. We had arrived with our product too late to be able to exploit it!

This innovation example is not only descriptive of transverse as opposed to lateral thinking but is an example of the interaction of invention and design as

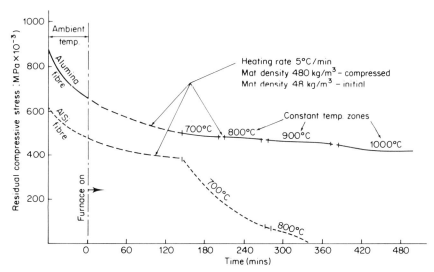

Figure 19 Monolith fibre support — thermal stress relaxation/time (Reproduced by permission of ICI Plc)

described by Gardiner and Rothwell (1985) who lay great emphasis on the role that design plays in the innovative process and in constituting a link between its various elements.

5.5 QUALITY CONTROL

So far, we have considered the importance of quality control of a product only very briefly, but it represents a vital element in the creation of any new product. In the previous section, several instances where such control was a vital factor in the applications and end forms concerned, have been highlighted, but taking a broader view, quality control relates to the feasibility, applications and development phases of the innovation process and ultimately forms the basis of production management in a full-scale production plant. Also in all research a 'control' is essential in order to ascertain the consequences of changes in product formulation, processing and application. Eventually quality control becomes the means of securing replication of quality, as well as monitoring tolerance standards. As such, it becomes the major tool of production control.

Controls used in the way described, introduce the experimenter to an awareness of the critical parameters governing 'properties' and 'effects'. They also signal to the innovator where the limits of the manufacturing process are likely to lie, in terms of achieving a consistent control by envisaged production methods and by laboratory methods of sample preparation.

Without the early inception of quality control a project is unlikely to survive when introduced to collaborators or potential users of the product. The point is that both researchers and purchasers must know what they are dealing with on receipt of samples or product. Without quality control on either production or research samples it is not possible to correlate performance of one sample with another, or to compare the new product with competitive products, nor is it possible to reproduce consistently the effects claimed for the product in a particular application or for that matter as a base product specification.

A valuable consequence of establishing quality control on all research samples is that it demonstrates to the recipient that the method used to produce the sample is capable of a measured degree of reproducibility of the properties and effects claimed for the product.

Such reproducibility lies at the heart of all mass production processes and leads us to the key factor of importance to a producer and user, which is; what are the tolerance limit requirements that quality control must achieve in selecting or processing a sample or product for release to the collaborators and users?

In setting quality control limits for a product we have to consider two interests: One is that of the producer who must ensure that the limits are economically attainable with his manufacturing process. The other is the user and especially the purchaser who is to utilize the new product in his own production line or application, where the limits demanded will be set by the peculiar requirements of the user's own process or application quality control requirements. Satisfying both these criteria requires careful consideration of the consequences of applying control limits on the economics of production of both product manufacturer and the consumer in addition to the technological demands that manufacturing to a limit or specification impose.

Clearly if the control limits are set tighter than the application strictly requires, it will lead to an uneconomic product cost, due to the higher reject rate which the producer will have to build into his costing of the product.

When the limit is set too low for the strictly defined requirements of the user, it may lead to a reduction in the new product cost, but because of the inadequacy of the limits which have been set, the user will inevitably experience unacceptable failure levels in his utilization of the product.

It is in everyone's interest to find the correct balance between limits and costs, in order that the overall business potential of both producer and user are maximized.

A quality control parameter needs to be one which can be easily and cheaply measured yet capable of indicating predictable changes in critical application performance, with which it must be correlated.

To illustrate the point, the monolith support system described in the preceding section, provides an example of the role of quality control in an application which merits further comment. In this project the base product, i.e. in the form of a 'fibre mat' was introduced into the exhaust system catalytic converter in order to support the refractory monolith by exerting radial forces on the monolith and the metal shell of the converter. The magnitude of the retention force achieved was controlled by the packing density of the fibre which was in turn dictated by the size of the annular cavity it had to fill. This cavity size was itself subject to manufacturing tolerances. The following details should clarify the principles involved:

From inspection of the respective requirements of the fibre manufacturer and the potential user it becomes clear that if either party fails to adhere to the limits imposed by quality control on their respective products or applications the consequences will immediately reflect on the other.

Working with one particular manufacturer in the USA we ran into this precise problem. It was due to a failure in communications between the two parties to ensure that the assembly tolerances of the user were compatible with the 'fibre mat' limits or tolerances. Perhaps a classic case of the innovator and the marketeer failing to 'get their act together'. The comments given earlier in this chapter regarding the interface between them, clearly have a practical relevance to ensuring exploitation success.

All quality control data is eventually summarized in the form of a 'research log' product data sheet on which basis preliminary product and process cost data must also be compiled. Obviously this form of data sheet is the precursor to the compilation of a market research and sales data sheet which will include selected parameters of the product with emphasis on its performance characteristics. It will exclude all know-how on processing and only expose formulations if patent publications have already revealed them. Even in the latter case a considerable degree of selection will be made in its preparation for publication.

Many new products that possess superior properties to their competitors can only be economically exploited if the enhanced performance they possess is fully utilized in the application, because this enhancement invariably is only achieved at increased cost of production. Simply aiming at a market because the product is 'better' is not the issue. What matters is that the product specification matches the utilization and only then can the economic argument begin to take effect. Cost-effectiveness is the criteria on which each product must ultimately be judged. Some products are, however, too sophisticated for the applications on which they are targeted and may fail the cost-effective assessment because their properties cannot be fully utilized technically in the application.

Methods for assessing cost-effectivenes can be developed for specific products which can then be applied to other products in a similar category. One example is in the assessment of load-bearing structural materials in which the strength of the material determines the unit mass needed for a specific duty related to its configuration and design, e.g. new materials which can be envisaged as structural members such as beams can be cost/effect assessed by setting out a standard loading configuration and geometry and then calculating the inertial cross-section of beam required to support the load at the recognized limiting stress for each material. By including unit mass/cost of the material in the calculation the relative cost of supporting the load in the stated geometry can be established, i.e. maximum cost/effectiveness can be established in relation to each candidate material. Further consideration is given in Section 6.4.

It is for reasons typically as the above that it becomes necessary to search for wider applications that might more fully exploit the base product's properties, even although this may entail further processing of the product to a new end form. The point is, will the secondary product which emerges provide a more cost effective utilization of the initial invention?

Another factor which has to be considered at this stage, is will the initial choice of product and application be able to achieve a sufficiently large market share to ensure production on an economic scale? If the answer to the latter is no, the case for commencing the search for new and more diverse applications becomes urgent for the project to survive.

The Innovative Project Model, indicates that applications research and market development, incorporating market research forms the bridge between the base product and the market requirement. This again highlights the critical nature of the liaison between the innovator or inventor and that of the marketeer.

5.6 APPLICATIONS PATENTS

As the applications phase progresses, the degree of secondary innovation will be realizing secondary products as we have seen in the earlier sections of this chapter. It is obvious that the same alertness to the inception of any applications inventions must apply as was evident with the base product. The important point is that examination of the idea, experiment interface is likely to make clear any inventive steps that can form the basis of a patent application.

Close contact with the Patent Agent is essential throughout the project and the innovator or inventor have the responsibility to take the initiative in bringing to his attention the information that they think may contain the inventive element. In the case of applications it is even more important to inform the Patent Agent at an early stage of the exposure of the product to the market in order to carry out joint evaluation as this makes it essential to secure product and process confidentiality, which is critically associated with the requirements of a patent application.

The illustrations already given of collaboration with potential customers for the product have shown that the embarkation on collaborative field trials must be carefully negotiated. Confidentiality is essential in order to be able to subsequently substantiate any claim to invention of the way in which the product is to be used to achieve effects or perform to meet other application requirements.

Confidentiality, know-how, patent application and strategy all receive further consideration in Chapter 7.

5.7 PROJECT REVIEW OF THE APPLICATIONS PHASE

By the time we reach the end of the applications research phase the base product will probably have been recharacterized and improved, applications products will have been introduced and market development for both base product and applications or secondary products will have reached the stage where any further advance demands that the 'product samples' are manufactured to consistent quality control standards.

At the review, the status of the project must receive highly critical examination, since the investment requirements in terms of research and market development, now starts to escalate rapidly.

The basis of the review is the data sheet, listing all the available technical data on the product including that data related to the application product as well as the base product. It soon becomes obvious that the product exploitation is revealing the 'front runners', i.e. those versions of the 'product' for which market interest is strong and which exhibits potential demand on a scale that could indicate that the production levels could result in a cost-competitive product.

In addition to the cost consideration, which increasingly dominates the project, the areas of market interest and the production potential it represents also has to be questioned in terms of; does this fit the existing business or merit toll-conversion? In the limit; do the answers to these questions point to the wisdom of dispensing with ideas of exploiting the product by manufacture in the innovating company in favour of selling off the know-how or patent rights to others?

Whichever option is selected it will be clear that once the appropriate exploitation strategy is adopted and products selected for priority exploitation through the market place, a new phase is precipitated in the innovation process. This is the development and design phase.

The final thought on the applications phase is that although the innovation project model postulates three distinctive phases, prior to embarking on any attempt at project exploitation, it also argues that these phases overlap. This means that while we talk in definitive terms of the end of a phase we have to recognize that this may not take place until an adjacent phase is itself well advanced. In fact, any attempt for us to place too tight a logic around the events which constitute the innovative chain can be counter-productive. If we look back to the earlier chapters we find that Rothwell has reminded us that innovation is a reiterative process. Beginnings and endings have a habit of prevelance in our innovative experience. *What I have attempted to do is to present information in what I trust is seen as a logical way for the sake of the reader's comprehension, but we cannot expect the innovative process to abide by our efforts to bring such tidiness to its practice! To resurrect our analogy with the visual artist, out of chaos comes completeness!*

6 Development and Design

6.1 THE ROLE OF DEVELOPMENT

In many instances the initial idea for a new venture project undergoes a significant transformation throughout the innovation process. We have already seen how ascertaining the feasibility of the idea, dominates the first phase of the process, during which time the idea is realized in the form of an experimental laboratory sample product. From the 'sample', the innovator can discover a wide range of information on its processing or manufacturing characteristics as well as details of properties etc. He can also identify novel 'effects' which may be attainable, when the laboratory sample is used in a variety of circumstances and in specific applications, often as part of a 'system'. This is particularly evident in the case of the innovation of 'new materials', where the identification and measurement of 'properties and effects' can be obtained from extremely small samples.

The latter point, i.e., of the usefulness of small samples, is particularly important to the lone inventor-entrepreneur working in such fields as new materials, as it means that he can advance his ideas through both the feasibility phase and the subsequent applications phase of his projects without having to resort to producing large samples which make much greater demands on resources in terms of technological and financial considerations. Other fields of research which offer similar advantages to the inventor-entrepreneur include microelectronics, where the scale and capital requirements are also amenable to the initiative of the lone inventor-entrepreneur.

Referring to the new materials type of project; examination and testing of samples can be particularly effective in establishing some idea of the product's potential for use in a variety of applications, using simulation techniques to assess the sample performance at critical operating conditions which model the application in question. An example of this is where a material is subjected to many compression loading cycles on a laboratory testing machine, in order to simulate the degree of compressive strain calculated to occur under the full-scale loading conditions when the product eventually goes into service, e.g. the compression may occur as a result of repeated expansion and contraction of a shell packed with the sample material and subjected to heating and cooling conditions, as indicated in the example given in the previous chapter (Fig. 19), relating to a catalytic converter monolith wrap fatigue test.

In many instances, the design of such simulation tests must be done in collaboration with potential customers for the product and this inevitably leads to the opening up of some aspect of the exploitation phase during the applications or development phases of the project.

As the project makes its transition into the field of applications research, the limitations as well as merits of the product are discovered. In particular the size and lack of sophistication of the laboratory samples used to advance the project to this stage is found in some instances to eventually impose constraints on its evaluation with potential users. It also becomes evident that the initial base product does not always prove to be capable of meeting the requirements of those applications which may have originally been identified as suitable targets for its exploitation. Development of the base product, allied to efforts to innovate secondary products to increase applications potential then becomes the focus of the innovation process at some point in the applications phase.

It can now be appreciated that development of the base product and the introduction of secondary products anticipates the start of a 'development activity' within the applications phase. This is portrayed in Fig. 3, as the overlapping of the application and development phases. Exploitation progress of both basic and secondary products may of course be limited by the fact that the laboratory samples, available in the application phase, are not sufficiently representative of the kind of full-scale production products envisaged and on occasions it may not be possible to effectively evaluate them by simulating service conditions in the laboratory.

The inadequacy of the laboratory sample in the eyes of the market is not only a concern that properties and effects may change when the product reaches the production stage, but that its capacity for meeting specific requirements, perhaps in appearance, form or other aspects which market exploitation demands, cannot usually be assessed from them. All of these requirements must feature in some form of quality control monitoring of the process of manufacture and use, which will anticipate the control system of the eventual full-scale type of process envisaged as discussed in the previous chapter. It is at this point that the project need is asserted for prototype product samples to support further development.

It now becomes evident that the development phase objective is, *the design and development of pre-production prototypes of the product, capable of meeting market specification requirements so that they can be used to establish or consolidate an entry into the market place, often by joint evaluation programmes with the potential customers for the product.* In the limit, new venture research and innovation is about establishing confidence in the ability to turn ideas into viable products which can be quality controlled and produced economically. The prototype is the means of demonstrating these parameters and an essential factor in confidence building.

In many projects, the level of production of prototype products can be met by a redesign of the laboratory sample processing or manufacturing facility, without the necessity for designing and constructing a more sophisticated small-scale manufacturing plant, i.e. a semi-technical plant, but as we see from the Innovation Project Model in Chapter 2, i.e. Fig. 6, depicting the development phase, process development interacts closely with product development. This implies that product specification will influence the method of production, while process specification has an impact on product definition.

Transforming the laboratory sample product into one which interprets all of these influences and maximizes its functionality as well as its aesthetic appeal, is met by the preparation of a product design that anticipates full-scale production methods for its manufacture, while meeting identified market test criteria. This represents a step change in the innovative process which can lead to a modification of product properties and performance, as well as changes in appearance. All of these factors characterize the prototype product and distinguish it from the laboratory samples on which product feasibility was established.

To summarize: the role of development is to advance the product to meet the market acceptance levels or specifications by applying the results of the feasibility and applications research phases, to the development of the new product, which until now has been represented only by laboratory samples. This developed form, we call the pre-production prototype, which must reflect contemporary views of the future full-scale product specification and its anticipated manufacturing process. The latter involves the establishment of the basic design parameters for the semi-technical and full-scale production plants, but not their implementation.

6.2 DESIGN AND INNOVATION

Development depends on both experimentation and design in order to advance the project to the stage where both product and process model the essential parameters of their full-scale counterpart. The design contribution introduces the use of logical methods of ascertaining the performance characteristics of the prototype and determining by analysis, how to optimize the various parameters involved to meet an identified application. High in importance is the need to achieve a cost/effective product and process while still leaving room for creativity to introduce new ways of resolving the problems which are encountered at this stage. Innovation as a result of experimentation is obviously still active in producing new ideas for effecting the translation of the laboratory sample into the pre-production prototype.

An example of design-related innovation is provided in Chapter 3, where in the resolution of a design problem concerning tall towers and wind loadings, an innovative design approach was made in one instance to meet the existing loading requirements, but was superseded by an innovative and inventive solution in which the loading conditions were changed, instead of being complied with, by disturbing the wind eddies that were the cause of structural resonance.

It follows from our consideration of the prototype that the laboratory sample preparation facilities are unlikely to have the flexibility needed to adequately 'model' the kind of process operations expected to apply on a full-scale production plant. As a result, a laboratory production unit must be designed specifically for the manufacture or processing of prototypes.

The unit concerned will probably employ the use of batch processing operations, but these can still be designed to 'model' in principle, key continuous processing or manufacturing operations that the full-scale plant will be expected to employ. Such a laboratory unit might be rapidly improvised to ensure early production of the prototypes. The alternative course of embarking on the engineering design of a much more sophisticated and expensive semi-technical plant would involve a delay in the introduction of the prototype to market research and development and may not be justified in any case until the project is further advanced.

The required production capacity of the laboratory units will be determined by the scale of any current and anticipated 'user evaluation programmes', which have been generated in the feasibility phase or more probably in the applications phase. Design of the unit for the base product prototype will be a development of the existing laboratory facility used in the feasibility phase, but governed by the need to model in critical areas the ultimate process plant operations expected to be used for full-scale processing or manufacture. In consequence it becomes necessary to initiate notional design studies for both the semi-technical plant and the full-scale plant, Events 26 and 27, so that the unit reflects this information and incorporates it into its design.

As commented on earlier, the applications phase introduces secondary products for which additional laboratory production units are required to produce pre-production prototype secondary product versions for evaluation studies. The design of these products generally introduces more technological development requirements which must be approached in exactly the same way as applied in the case of the initial, or base product.

Reference to Chapter 5, Fig. 6, will provide a reminder of the dependency placed on the base product by the secondary products in the case of the innovation of end forms with which to meet new applications in a prototype product form.

The data furnished from the contributions of Events 22, 23, 26 and 27, ensures that the prototypes are likely to be a true reflection of the eventual full-scale production product, but also of importance, these events will provide the vital product/process costs relevant to both semi-technical plant and full-scale plant, completing the pattern of information on which the project exploitation strategy must eventually depend.

A semi-technical plant stage would almost always entail a change from laboratory batch type operations to those which are of a continuous production processing kind, usually associated with full-scale production plant. It is at this point that the technology of process design takes over from what we affectionately call the 'bucket and spade' stage of processing.

Clearly the resources required for the design and construction of the laboratory process unit will be much less than would be needed to make the leap from laboratory experimental sample preparation, as used for the feasibility phase, directly to the semi-technical scale for production of prototypes, but it must still reflect the continuous manufacturing process of a full-scale plant, although modelling it, often in several discrete batch type operations, which may be more suited to laboratory production or as dictated by other circumstances, such as cost and expediency etc.

The laboratory processing unit also provides an opportunity to investigate at the laboratory scale those process parameters which determine the design specification for such a semi-technical plant. Only if the research and technological assessment of the project eventually merits it and market research allied to applications research endorse it, should the question of jumping from laboratory scale to semi-technical scale production be given serious consideration.

The development and design phase model (Fig. 6), concentrates attention on those aspects of the innovative process that stop short of the actual design and operation of the semi-technical plant and the full-scale commercial production plant, *since the establishment of the Laboratory Process Unit alone, can in many cases be an adequate basis on which to attempt to secure the objective of exploiting the product and process concept as a new business venture.*

The lone inventor-entrepreneur will find that the Laboratory Process Unit approach is particularly attractive to him, as it will probably lie within his own resources to implement it and enable him to work through the prototype product development and design. In pursuing this course the innovator will greatly increase his bargaining position with collaborators or potential licensees, when seeking to exploit the project, with a view to persuading them to undertake the semi-technical stage of the project development and the ultimate design, construction and operation of a commercial scale plant.

If the prototype proves too sophisticated for it to be developed with the aid of the relatively simple laboratory production unit described, it may indicate to the lone inventor-entrepreneur, that his only opportunity for realizing a return on his innovative efforts will be to offer his ideas and/or patents to others to develop, on suitable terms, as discussed in the concluding chapter. On making this statement we are reminded of the often quoted criticism of the UK, i.e. that we are good at inventing but not at realizing successful business ventures out of them. The circumstances I have described may indicate that my earlier plea for a reappraisal of the adequacy of support for helping the lone inventor to make the transition into the prototype product should be considered seriously. The question might be asked, how do inventor-entrepreneurs fare in the other leading industrial nations of the world?

The research project team of the large company may find this laboratory 'step-by-step' approach to a project meets their needs, by providing scope for experimentation and development of both process and product as well as preparing the ground for the work they will later undertake with respect to the specification of the semi-technical plant and its operation.

6.3 DESIGN

The difference between the laboratory experimental sample product and the pre-production prototype of the laboratory process unit is that the laboratory sample serves to demonstrate the feasibility of achieving specific 'properties' or 'effects' inherent with the product, whereas the prototype product interprets the product in 'design' terms and in relation to production and manufacturing operations while also relating 'design' to specific application requirements. The prototype therefore embodies the 'properties' and 'effects' potential of the initial laboratory product in a form designed, produced and eventually submitted for market sampling purposes in which it is targeted on specific applications.

An example of a secondary product which is in this category is provided by the alumina fibre project referred to earlier. It consisted of an application for the fibre in the role of a filtration media for a wide range of applications.

The research team had perceived that a refractory fibre mat, used as a filter media had particular advantages if constructed from 'Saffil' alumina fibres, which are resilient and capable of withstanding high temperatures. They are also extremely fine fibres, which can present a high specific surface area as a particulate collection surface. One application stood out over all others and this was in the diesel engine exhaust emission field.

The motivation for this choice came as a result of the activity of the EPA (Environmental Protection Agency) in the USA, who were introducing at the time, new controls on the limits of particulate emissions from diesel engine exhaust systems on cars and trucks. This required the removal of fine carbon particles, or soot, in the range from submicron to a hundred microns or more in size as encountered in diesel engine exhaust gases. Microfine fibres were evidently candidates for such a task.

The principal function of the proposed filter was to remove the carbon particles which also carried a variety of obnoxious compounds on them in the form of exhaust smoke emissions. It was appreciated that any particulate filter would progressively lead to the build-up of back pressures on the engine system and to minimize this, the idea was introduced that the soot particles should be collected on the filter mat and then subsequently 'burnt-off' to regenerate the filter media, using a suitable catalyst to initiate light-off. In this way, it was hoped that the problems of rapid blocking of filters of this type could be avoided and in consequence engine performance could then be kept at optimum efficiency.

The analysis of the problem provided a basis for the design of a 'low-capacity' filter, of an acceptable size for use with an automobile, on the principle that the concept provided for regenerating the filter mat, by periodic high temperature peaks which occur in the exhaust gas emissions under normal service conditions and which would ensure that combustion of the particulates on the filter media would be achieved because the ignition temperature for the deposits was lowered by the catalyst.

The reason for the emergence of this potential market was as stated previously, due to the initiative of the Environmental Protection Agency of the USA, backed by the legislative powers of the USA Government. The result was that the automobile industry was set a target date by which the emissions from vehicles had to be reduced to a specific level.

This legislative introduction of technical performance requirements before the technology has been fully developed represents an important feature of how the USA gets things done. By contrast, the UK draws up specifications to meet identified product requirements, generally after the products have reached the market, i.e. British Standards Association activity is of this form, but compliance with these Standards is not generally enforced by legislation, but may be supported by Codes of Practice which set out recommendations for their application.

Where products are in use which affect the general public welfare, particularly in matters of safety, legislation may be introduced which draws upon the Standards and Codes of Practice to enforce specific limits on design, manufacture and use of nominated products, but as already pointed out this is only done after market experience has indicated the need for such legislative action, i.e., the legislative procedure is not used as a spur to innovation as in the USA model.

The latter philosophy has certain drawbacks for the innovators and entrepreneurs since ideas for product improvement often fail to obtain venture capital support, because it is considered by the business community that in the absence of legislation the 'customer' will not pay a price premium for the product which may incorporate desirable safety features. The USA approach is calculated to spur innovation to a much greater degree than is the case with the UK approach.

It is obvious that with a market as large as the USA 'truck' and 'diesel automobile', such legislation assures a ready market as a target for the innovator. It also inevitably generates much competition, since the degree of research investment into the innovation of a new product to meet the identifiable scale of the potential demand reveals the level of financial return attainable, against the investment, if the project succeeds.

The initial response to the challenge we now faced was to interpret the initial idea or concept in the form of a conventional 'through-flow' filter, in which the 'element' was a mat of 'Saffil' alumina fibre, i.e., in the base product form in which it left the production unit. On testing the filter it rapidly became evident that the rate at which the filter blocked was excessive and it was recognized that

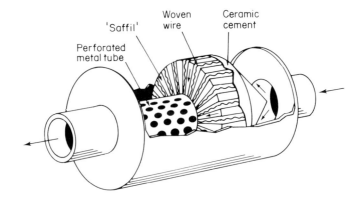

Figure 20 Alumina fibre soot filter for diesel exhaust systems — notional sectional drawing (Reproduced by permission of ICI Plc)

even with the introduction of a soot burning catalyst, the frequency with which the exhaust gases would reach the ignition temperature for the burn-off of the soot particles was going to be too infrequent. The alternative was to make a larger filter — a suggestion rapidly rejected — or to develop a filter with an element that did not block so rapidly, but which still retained the particulates.

Fig. 20, shows the design of a filter we developed, incorporating the base product of alumina fibre which was eminently suitable for the thermal and dynamic loading conditions to be met in service. Its feature was a new filter element concept, which aimed to reduce the rate of build-up of back-pressure on the system by adopting a 'diffusion' model for filtration to replace the 'through-flow' concept of the initial design. Fig. 21, shows the basic cross-section of the filter element, which consisted of alternating layers of 'Saffil' alumina fibre mat and crimped woven wire convoluted mesh. The identification of an appropriate catalyst completed the selection of the materials requirements of the project.

This particular example helps us to see how design plays its part in the innovation process. In the case of the element, an appreciation of the gas-flow behaviour and particle movement contributed to the idea of 'losing particles on the converging fibre faces of the labyrinth', or in more orthodox engineering terms, of achieving flow conditions in which the migration of the carbon particulates could take place as Stokes' law predicts.

It is evident that while the design contribution was important, the idea for the labyrinth did not originate from such logical engineering considerations which are in fact an explanation of how the filter media functions. Instead, the idea was the result of an attempt to prevent fibre loss from the mat by sandwiching it between crimped woven wire mesh, which transversely suggested that lower pressure drop might ensue as a result of the increased surface area of fibre mat presented to the gas stream.

Why woven wire? Why crimped wire? These are questions whose answers

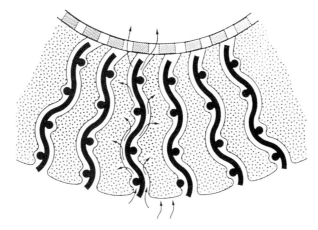

Figure 21 Diesel exhaust filter — element detail (Reproduced by permission of ICI Plc)

lead to the root of invention. The diesel filter project was taking place in the same environment as the work on the automobile catalytic converter monolith support system, referred to earlier, which employed the same alumina fibre and woven wire rings in the first efforts to restrain the monolith movement in the petroleum engine exhaust systems.

The transverse thought process was evidently at work, intuitively relating woven wire mesh and fibre restraint to a new environment for its application in association with an idea for providing a 'low pressure drop path' for the diesel exhaust gas stream in the 'soot filter', which involved similar dynamic loading conditions already encountered and resolved in the catalytic converter.

Behind this secondary idea was the thought that the reduced pressure drop could extent filter life and if this was prolonged enough, it would ensure that the incidence of high temperature peaks in the exhaust gases might give rise to 'burn-off' conditions with sufficient frequency to prevent the filter pressure drop from rising to unacceptable levels, as the soot particles were collected during service conditions. This concept was to form the basis of a prototype design.

Some aspects of the innovation of the filter can be seen as surprising and therefore inventive, while other aspects are recognized as the result of designing-out limitations in the filter media, so that we conclude that the innovation incorporated both design and invention in its concept.

The development aspects of this project lay in the transition from a simple drum type of filter element, employing conventional gas flow through the element, to the creation of the labyrinth filter element which depended on turbulence in a labyrinth, allied to gas diffusion and impingement to deposit the particles of soot on to the fibres without incurring rapid increase in pressure drop. A process of problem identification and in this particular case, of resolution by innovative and inventive ideas in which design played an important part but not the dominant part.

132

(Side View)

(End View)

Figure 22 Alumina fibre diesel exhaust filter — elliptical prototype (Reproduced by permission of ICI Plc)

The focus of the development phase is generally aimed at the production of a prototype that potential customers can immediately recognize as suitable for mass production methods of manufacture, but which also possesses design features that will fit in with environment in which the product will be deployed. In the case of the diesel filter, in attempting to introduce its use to cars in addition to large 'trucks', it was necessary to change from a circular filter as indicated in Fig. 20, to an elliptical filter, the core of which is shown in Fig. 22. This enables the filter to satisfy the required ground clearances, since exhaust systems have to be placed underneath the cars, whereas in the case of 'trucks' they are often mounted vertically behind the driver's cab, where a wider range of configurations can be tolerated.

As we have seen, development and design act in synergy in many innovations, as an entirely different type of project now illustrates. I refer to the way in which an 'inorganic bead or prill foam' resulted as a consequence of resolving a processing problem when attempting to make an aqueous based inorganic foam by a process in which the foam slurry was cast to produce rigid foam slabs or boards. The problem encountered with the casting process was one of shrinkage and associated cracking of the foam during the drying operations.

The foam had all the appearance of a clay river-bed in mid-summer — not in England of course — craze cracked! I had been busy 'throwing' clay pots in my spare time and had learned that the best way to reduce the incidence of shrinkage, which occurs on drying the pots to the 'leather hard' state, was to mix 'grog', i.e. fired particles of clay, to the wet clay before 'throwing'. What relevance was this?

Just suppose that dried foam particles were to be added to the wet continuous foam, might this also reduce shrinkage on drying as in the case of the pottery clays? In the event, the process team went on to experiment with dried inorganic foam beads or prills combined with the aqueous foam feedstock to achieve an integrated product which did not crack on drying during press moulding of the boards.

This development was not exactly what I had anticipated as the proportion of the mass of 'beads or prills' to the amount of feedstock foam slurry was very high and resulted in the matrix being formed almost entirely of beads, bonded together with the slurry material. This meant that the resulting boards were somewhat friable. My initial ideas on the subject were that the beads should sit in a continuous and homogeneous matrix of aqueous foam in a way more comparable with the clay analogy cited earlier. My guess is that, friability would have been reduced if the latter construction had been adopted.

What really mattered at this stage was that the cracking problem could be resolved to a level of acceptability to meet the expected application requirements.

Design had an important role to play in ensuring that the proportion of the boards that ensued, could support their own weight over spans normally associated with the use of a product that clearly was suited for the building industry as roof or wall insulation etc., where its inorganic fire resistant construction could be exploited. This meant that the thickness of board required to support itself over a range of 'standard' panel sizes encountered in roof or ceiling applications had to be calculated, based on the strength data determined during the feasibility phase of this secondary product.

The importance of this kind of preliminary design study is that it sets a target for the developer to satisfy, i.e. developing and fabricating the product in a board form to predefined dimensions and tolerances. We have arrived again at our old friend quality control in order to satisfy product design specifications and exploitation criteria.

6.4 COST/EFFECT COMPARISONS

Design plays an important part in ascertaining the potential cost/effectiveness of a product and its competitiveness with other products already on the market. Having reached the development and design phase and the prototype design stage in particular, we have at our command sufficient reliable data on our new product to enable us to revise any preliminary cost estimates for the process and product and then apply these when making the cost/effectiveness assessment.

To illustrate this aspect of innovation in a little more detail I refer to a project in which I played a small supportive role. It concerned the innovation of a high strength inorganic cement which was capable of replicating the type of forms and performance we normally associate with organic materials, e.g. the thermoplastics and thermosets. The idea behind it originated with Professor J.D. Birchall FRS (1983a), who has emphasized that the abundance of the inorganic materials throughout the world could be used to effectively displace many of the 'plastic' products available, providing inorganic material could be made which exhibited a similar facility for mass production and possessed comparable physical properties, in particular those of strength and toughness. Such a transition could then be expected to lead to a conservation of fossil resources, which form the basis for the manufacture of the 'plastic' materials. This in turn would help in conserving world energy resources as represented by the fossil fuels.

As a result of this philosophy the project had a clear objective to 'displace' other materials which possessed specific attributes, but in particular to target on an inorganic material which could in the limit displace the dependency on organic materials whose origins were found in feedstocks which represented sources of fuel. A key part of achieving this target depended on monitoring any progress made in the project towards attaining the desired physical properties in the new product by comparing its cost/effectiveness with competitive materials, using some simple and convenient measure of strength in relation to mass and unit mass/cost for given configurations as introduced in Section 5.5.

To explain in a little more detail; whenever we think of a material which claims strength as one of its attributes, we think of its ability to support itself over spans which might be encountered in a variety of applications that relate to the forms and size of product which we envisage will be attainable from a fully developed commercial scale plant. The building industry is a typical example which allows us an opportunity to make direct comparisons with other materials which are utilized in a similar way.

A widely used basic form of construction member encountered in the building industry is the beam. Cost/effect, comparisons can be made using this as the 'model' form. A relationship between tensile strength, span and moment of inertia of a beam constructed from a material whose unit mass cost is known, can be established to produce a value for cost/effectiveness and applied to a variety of materials for which data is available. Comparison of these calculated values makes clear which material in the chosen form and application, will provide the most cost/effective product.

Not all materials lend themselves to economic production in the same form, e.g. a tube of metal might be extruded, but a beam of wood is in general cut into a rectilinear shape, yet both could be offered as candidates for a structural use as a beam.

With this diversity in mind, such variations are catered for in the use of the moment of inertia in assessing the relative merits of the materials, since the moment of inertia describes the cross-section geometry of the beam precisely.

Cost/effectiveness sounds like an excellent basis on which to judge any product, but it is essential to appreciate that the conditions under which such evaluation is applied can differ according to external factors, as the following anecdote shows:

It was a memorable occasion when presenting the results of some measurements of the 'properties' of one of our new products to a leading automobile manufacturer in Europe. The results of comparative testing showed that the product was ideally suited for the application concerned and what was more advantageous, from the exploitation point of view, was that the comparison revealed that the competitive product was operating at the threshold limit of failure under the loading conditions imposed.

We might imagine that the introduction of our product to his production line was inevitable, but the manufacturer was faced with gaining product approval from statutory bodies which involved lengthy and costly retesting of his system which had already gained approval when specifying the competitive product.

It was not to be. The manufacturer produced statistics that indicated to him that it would be more costly to change to our product than to face the relatively small number of failure claims he could expect to meet by retaining his existing system using the competitive product.

Not even superiority of performance, combined with competitive product cost, i.e. cost/effective superiority can overcome the strategic decisions that sometimes arise during the development and exploitation of a new product. The criteria of successful development and eventual exploitation is that the product provides the customer with an overall technological/cost advantage when assessed in a contemporary business situations.

On the occasion in question, I countered with the comment that 'the decision to retain the use of the existing product was taking risks on the predictable level of failures'. Politely the companies' representatives agreed, thanked us for our research and our 'presentation' and held to their decision. It was time to pack and head for home with a technological success, a sound cost/effect analysis supporting our claims for its superiority — and an exploitation failure in our pockets!

Exploitation of the product and ultimately of the project go hand in hand. Development prepares the ground for the exploitation phase.

6.5 PROJECT TECHNO-COMMERCIAL ASSESSMENT

As we reach the advanced stages of product development we begin to draw together all the mass of information that characterizes the project, much of which will already reside in the form of the product data sheet. This will chiefly consist of properties and effects, but this information must be supplemented by a review of the status of the applications which have been opened up in the applications research and development phases.

Taking stock of what our project has to offer a prospective client, is now a deciding factor in the timing of any move to initiate the exploitation phase objective of securing internal capital investment decisions with a view to eventual production of the new product, or of selling/leasing, manufacturing know-how and/or patent property to others, to turn the project into a manufacturing business new venture enterprise.

What any exploitation strategy must be based on, is a sound presentation proposal to put before those we seek to interest in the project. This represents the overlap of the development phase with the exploitation phase as Fig. 3 indicates and must be largely compiled in conjunction with a critical review of the development phase, as the concluding event of Fig. 6.

The content of the presentation proposal is discussed in Chapter 7.

7 The Business Objective – Product Exploitation

7.1 EXPLOITATION STRATEGY

The exploitation phase marks the culmination of the innovation process chain and is represented by Fig. 7, as shown in Chapter 2. In starting this phase we find that the results of all the innovative activity of the feasibility, applications and development/design phases, have been drawn together in the compilation of the product data sheet, which has been described in Chapter 6. The chief constituents of the data sheet, are the 'properties' and 'effects' relating to the product and these eventually form the 'core information' for a project exploitation proposal, when supplemented by the notional process and product cost studies.

The ultimate objective of compiling this exploitation proposal, is to be in a position to present a concise outline of all the key features of the project to business managements or entrepreneurs, in order to secure their adoption of the project with a view to ensuring that it forms the basis of a new business venture. For the team innovator or the lone inventor-entrepreneur, this represents the key target of their endeavour, but in the entrepreneur's case it provides the opportunity for establishing a financial return on his research as early as possible. Obviously the entrepreneur has the alternative course open to him of raising his own capital for the design and construction of manufacturing plant, followed by production and marketing.

These latter steps often demand skills which the inventor-entrepreneur may not possess and resources that he cannot command, for that reason, I have focused attention on making the exploitation target the selling of ideas, know-how and option rights on patent property as a priority when presenting the proposal. The sequel to such a strategy is of course to establish a Licensing Agreement on which a financial return in the form of a percentage on product sales can be secured, should the *option period* convince the client of the technological and economic future of the project.

In planning our exploitation strategy we will have made some preliminary decisions following the initial feasibility phase, i.e. at the start of the applications phase, but our appreciation of the project will have matured by the time we reach the end of the development phase and the contemporary situation we face may now be much different from that with which we started. Even changes in personal contacts within our own organization or collaborators may influence our ideas at this later date.

What we have to consider is much more than the 'properties' and 'effects' of the product. Its market fit and its applications potential are of particular importance in determining where we can expect to find a client and in demonstrating the potential requirement for the product. This also leads us to consider the nature and degree of competition the product might expect to face in the market place and to make an assessment of the client's technological and business resources. In addition it enables us to form some idea of his business interests, with particular regard to how a new venture based on the project would fit in with them. Of course much of this work will have been carried out at the applications phase, as earlier examples have shown, but it is important to note that while such questions as — how does the potential client's balance sheet look? — will have been the concern of the marketing and business management personnel of the corporate venture, the lone entrepreneur must extend his own assessment of the partnerships he seeks in the same way and with even more awareness of the consequences of making the wrong decision.

Pertinent to all of these considerations, questions of patenting versus know-how arise which are central to any exploitation strategy and as later comments show, these can determine the 'commercial value' of the project to the innovator. Again these aspects of patent strategy will have already been reviewed when drafting the initial patent application at the feasibility phase, but as the project has progressed, secondary invention often occurs and vital information is generated in support of the patent application. As a result the possible requirement for exerting 'confidentiality constraints' over some of the information relevant to patenting and know-how becomes an important issue which must be reassessed before attempting to implement the exploitation of the project by negotiating option rights etc. This aspect will also have been kept under review throughout the innovative process, especially when first considering any contact with collaborating organizations or potential clients.

A number of exploitation options are open to the innovator, but each needs to be evaluated and priorities identified in order that the presentation 'pitch', can attempt to secure the most appropriate 'deal' in the subsequent negotiations to secure terms on which the project may be transferred to a client. Equally important is the need to identify the reserve exploitation option if the priority option cannot be implemented. This ensures that the initiative and movement towards project adoption can be maintained, once the negotiations have been opened.

This broad introduction to the exploitation phase will I hope have served to indicate that the development of a sound exploitation strategy must have its roots in the early stages of the project, i.e. it must govern the number and type of contacts to be established in the market place when seeking collaborators to assist in product evaluation, generally under some form of confidentiality terms. These contacts are then likely to lead to identifying a potential client to whom the exploitation proposal will eventually be addressed.

From the above we see that the exploitation strategy is highly interactive with all the phases of the project, but culminates in securing the future of the project in the hands of a suitable client while at the same time realizing a financial return on the venture for the innovating company or lone inventor-entrepreneur.

7.2 CONFIDENTIALITY, KNOW-HOW AND PRIOR-ART

Many new venture projects are characterized by an attempt to create both a new product and a new process in association with it, which as the previous chapters have shown, depends upon the generation of ideas, invention and know-how relating to product or process. These three elements are of fundamental commercial value to both manufacturer or business entrepreneur because they constitute the core of essential information on the product's properties, guidance on the means of manufacturing it, indications of its application potential and of secondary products, as well as being the repository of data on market opportunities.

A manufacturer or business entrepreneur is most likely to be interested in acquiring the patent rights and/or know-how to manufacture or otherwise exploit such a product, providing the exploitation proposal presentation made to him contains convincing information of the product's business potential. In addition he will be attracted by the fact that key aspects of the project have not been disclosed to the public at large. His interest in the latter is to avoid competition during the inception period of marketing the product where any premature disclosure of key information on the project can destroy the commercial value that it holds for him and lead to a rapid termination of any prospect of the innovator achieving his own exploitation target of securing a buyer for his know-how and inventions. In particular, premature disclosure of the content of a project, i.e. before patent applications have been filed, will invalidate any prospect of being granted a patent.

Ensuring the 'confidentiality' of the project is therefore an important part of successful product exploitation. The extent to which we implement a policy of internal confidentiality will vary according to the project status at the time, e.g., if patent applications have reached a publication stage, there is little point in demanding close confidentiality on those aspects of the project already protected by it, but not all of the project information we wish to protect is necessarily included, since we often encounter some aspects of the process or product that are not novel and must be considered as prior-art. It follows that we define our application of such knowledge as know-how, whose confidentiality must be secured, because we have no other way of protecting our ideas or how we achieve their implementation in the case described.

Confidentiality has however to be exercised at two levels of activity; the first is among the participants in the project, i.e. where it is a case of — do not discuss the project outside the team or company or communicate its details in any way to others until authorized to do so. The second case complements this policy and refers to those we wish to make confidential disclosures to, i.e. in formal negotiations with collaborators or potential customers involving everything from preliminary field trials and evaluation programmes to the more critical kind of negotiation where the offer to grant patent rights and know-how is to be made as part of a business arrangement which represents on many occasions the exploitation target for the project.

In achieving internal security of information, we depend on good management and trust to play their part, but most large companies insist on employees signing conditions of employment that include the requirement that they will maintain the confidentiality of the companies' business, technology and know-how.

In the case of external security it is usual for the innovators or inventor-entrepreneur to prepare, or have prepared by the Patent Agent, a Confidentiality Agreement which is submitted to the collaborators or potential purchasers of the patent rights and know-how of the project, for counter-signature before embarking on any discussion relating to the project. Failure to take this precaution can represent disclosure of the know-how, no part or whole of which could then be subsequently patented, assuming that it possessed the necessary novelty which is required for a patent to be granted, as indicated earlier.

It will be obvious by now that confidentiality is of crucial importance in the innovation of new products, but it might be thought that if a patent application has been filed, such precautions as confidentiality agreements are unnecessary. This is not always the case, since the application for a patent is not published until one year has elapsed. Disclosure during this period represents the release of privileged information which may adversely effect the exploitation prospects for the project. In other words it alerts the competition to the project with the result that although they may not possess the full details of its content, they will be able to deduce the probability of it having specific characteristics and initiate ways of countering any advantage they anticipate the product may have over their own product.

Know-how is a valuable part of any exploitation package, embracing both technical and commercial factors. It may in fact include inventions, for which the strategy adopted, determines that no patent application should be made, but which must consequently be protected by a confidentiality status if possible. The reason behind this attitude, i.e. of not always applying for patents is that the patent discloses in sufficient detail, how the product may be produced and provides information on such items as basic properties etc. Such information can often be used by competitors, to initiate a programme to reproduce the innovator's manufacturing process, familiarize himself with the features of the project and then identify areas for improvement or redesign.

From this position of awareness the competitor can proceed to a research and development programme aimed at achieving the improvements desired in the product or process with a view to securing patents or know-how of his own, in order to protect his existing markets, perhaps defeating our capability of exploiting our new product in them. The competitive edge is sharp and keeps us ever vigilant. The outcome, is that the most cost/effective product will be expected to triumph.

This emphasis on confidentiality is not of course confined solely to patenting considerations. Many aspects of a project are in effect prior-art, which we can define in the context of product innovation as methods of performing specific manufacturing operations which are already public knowledge.

Where such prior-art represents a critical step in achieving the cost/effectiveness of the product or process of manufacture its importance is elevated to that of 'know-how' and we shall then wish to keep secret the fact that we employ it to achieve our product. In this way we can minimize the chance of the competitor replicating our process and product by denying him access to selected areas of our knowledge.

Confidentiality then, assumes its most significant place in the exploitation strategy when we enter the exploitation proposal stage with external clients who may typically be manufacturers, developers or business entrepreneurs. In such cases, formal confidentiality agreements must be drawn up between the parties concerned with respect to any aspect of the project which the presenter or his advisers consider would prejudice its commercial value if published or used by others without consent. This applies in particular, to information which would form the basis of any patent application, but clearly includes know-how.

The alternative to the confidentiality based proposal is that selected areas of confidentiality may be excluded from an initial presentation. This is more practical than appears at first sight, since it is possible to make wide disclosures on 'properties' and attainable 'effects' inherent to the product without demonstrating the 'how' of achieving them. It is the 'process of manufacture' which is often the subject of know-how and patent application, but the performance and characteristics as represented by 'properties' and 'effects' which are of first importance to the client in the initial stages of any consideration of project exploitation. Disclosure of the latter alone may serve to establish the credibility of the proposal for the client, without destroying the inventor's bargaining position or threatening his patent application strategy, but a consultation with a patent agent is obviously highly desirable when formulating strategy.

Questions of 'how', represents a separate stage which becomes an essential topic for discussion if the potential client recognizes from what he has learned, by the way of 'performance' of the product, that the project now merits further considerations, especially regarding what is entailed for him if he were to contemplate investing in the product's exploitation by furthering its manufacture. In that event, a formal confidentiality agreement would be essential before a more comprehensive presentation of the product and its processing could be initiated.

The necessity for confidentiality in relation to patents and exploitation strategies can be seen to be of paramount importance in particular for the lone inventor-entrepreneur for whom know-how and/or patent property constitutes the major element of his equity. It is also to be noted that these two properties also provide the corporate innovation team with an important exploitation option to embarking on further in-house development of the project.

Occasionally we fail to appreciate what information is of critical importance to a product exploitation, with the inevitable result that such information is imparted to competitors who then quickly take advantage of what they consider to be prior-art, in the absence of any confidentiality agreements.

One such instance occurred when 'presenting' a project to a company who appeared to be potential clients for our work. What we did not know at the time was that the company in question were themselves already engaged on research into the same product field. The result was that apparently harmless disclosures — on a test method we had adapted to simulate the service conditions demanded of our product in a specific application — appeared in retrospect to have been carefully noted and incorporated into the test procedures of our hosts. The test was an important means of identifying the degree of variation in the product made on the pre-full-scale process plant and hence a means of achieving quality control sampling of the product. It was therefore helpful in demonstrating the product's suitability for the application in question and critical to its exploitation. Some weeks later we saw publications on the performance of the competitive product which illustrated its properties by the use of the type of test we had identified for our own research. We should not have disclosed the nature of our test and instead should have kept to the results of field testing, with the rider that we had a method for assessing quality control by a product sample procedure before assembly of the product in the user's application. Another failure — another lesson!

7.3 PATENTING AND LICENSING

Patents or patent applications are as already pointed out a vital part of the equity of the inventor-entrepreneur and for this reason must be carefully considered in the development of the project. We have already discussed that patents will only be granted where novelty can be demonstrated and shown to possess an element of surprise. This distinguishes invention from design in which a logical utilization of knowledge or prior-art is pursued to produce a product which can be anticipated. Fig. 11, in Chapter 4, shows the interactivity of idea, and experiment in relation to invention and study of this chapter will assist in determining if the innovative project under consideration has yielded invention which meets the definition given.

On the assumption that invention has been achieved, the procedure for making a patent application in the UK is laid down by the Patent Office, but it has to be appreciated that once an application is filed it must be made public after an eighteen month period has elapsed. After this time the information it contains is freely accessible to any enquirer. In other words the competitors can learn the essential details of the product and/or process and perhaps improve on it with ideas and inventions of their own. The first lesson we have to learn is that while early application for a patent obtains an early priority date for the invention contained, it has to be weighed against the disadvantage that early disclosure may be embarrassing if the project cannot be advanced to an exploitation stage in a reasonable time.

Patents applications may need to be filed in a number of different areas of the world, which will depend on where we anticipate the product may be manufactured. Typically a UK application can be extended to cover all the countries of the EEC, at an increased fee, to cater for the possibility that one of its member countries may prove to have the major market potential for the new product, and therefore provide the logical location for the manufacturing plant. Alternatively the USA or Japan might be the relevant area for filing applications, which raises a different set of circumstances for us to consider.

In the case of pursuing the applications outside the EEC it must be pointed out that the process of securing acceptance can be lengthy and costly, as it can involve the necessity of employing an attorney to progress the application with a Patent Agent and the appropriate patent authorities. Overseas filing has accordingly to be only undertaken with the awareness of the probability of encounterng a rapid escalation in application costs.

To the lone inventor-entrepreneur, the patent application cost issue presents him with serious problems, whereas the team innovator will be aware that his research budget fully anticipates such costs and his own organization will have its own patent officers or alternative contractural arrangements.

At the feasibility phase the question that faces the lone inventor-entrepreneur is — what do I do about patenting any inventions I produce? — His next question is — do I attempt to file my own patent application or use the services of a Patent Agent? Cost considerations come immediately to mind since an individual inventor, with perhaps no assured income from his endeavours, can find a Patent Agent's fees to be an expensive business, but in fairness, the Patent Agent does not usually constitute the major cost of the patenting operation.

The basic cost of patenting in the UK breaks down into agent's fees, HM Patent Office application fees and patent maintenance fees, but once the applications reach out to secure patents in the EEC, the USA and other areas of the world, the cost of fees increases and an additional cost for an overseas attorney to provide support in the pursuance of disputed claims may also arise. This latter point is often met as a result of the interpretation placed on claims by different patenting authorities and countries.

Guidance on patenting for small firms in the UK is available on a business consultancy basis from the various Small Firms' Advisory Units, supported by the Department of Trade, which provide the first two consultancy's free and offer to assist in an initial patent application whose aim is to secure a 'priority date'.

The UK Patent Office themselves publish guidance for inventors, but where a project is likely to involve a process and product patent I consider that it is advisable to enlist the personal aid of a patent agent, as he can offer specific advice on the potential of the application and often clarify exactly what has been invented. In the latter respect I would refer the reader to Chapter 4, where the relationship between idea, experiment and invention is discussed, and illustrated by Fig. 10 where the form of the interactions which take place at the inception of innovation serves to clarify if, where and how invention arises and this should assist in the preparation of patent claims or in briefing the Patent Agent on the course of the innovation.

It is as well at this stage to point out that while some innovative projects may not qualify for patents and retention of confidentiality also prove impracticable, they may be candidates for protection in the form of a Registered Design, published by the HM Patent Office. The essential point here is that products that depend on appearance or a design which is based on prior-art, instead of novelty, qualifies for Registered Design status, but not for a patent.

7.4 DATA SHEETS AND EXPLOITATION PROPOSALS

Throughout the innovative process, technical and marketing data is gradually built up to form a significant package of know-how on the base product and any secondary products. Using the innovation of 'new materials' as an example such data will typically include the following information:

(1) Description of product or products.
(2) Properties and quality control tolerances.
(3) Effects attainable with products in specific applications.
(4) Amenability of products to 'forming' and end processing etc.
(5) Application potential.
(6) Compliance with statutory regulations.
(7) Compliance with non-statutory codes and specifications etc.
(8) Market competition.
(9) Notional cost estimates of products and processes.
(10) Process of manufacture.

The inventory of data at this stage is in effect a summary of the research log-book in which is compiled information on the evolution of the project. This is supplemented by the data obtained from market research and preliminary business studies. This enables the entire range of exploitation possibilities to be identified. Such data is usually the subject of a project review where the ongoing status of the project is kept under constant scrutiny.

From this data, 'selected information' is extracted to make up a product data sheet for publication to potential purchasers of the product or to those who will be considering the purchase of patent rights. Care has to be taken not to include information which should instead be kept as know-how as discussed in Section 7.2.

This brings us to the consideration of how we intend to use the information on the data sheet when seeking to exploit the project as a 'business package'.

The data sheet will form the basis of any exploitation proposal which entails the careful compilation of a project presentation to potential users or those interested in establishing a business venture on the basis of the new product through manufacturing or acquiring patent rights, but it will be found that much greater interest is gained during a presentation if the data can be backed-up with product samples including those that have been subjected to a variety of tests in which the parties concerned display a vested interest. In particular, results of those tests in which the collaborators have participated or contributed in some way.

The tests referred to differ from those of the product research stage where basic 'properties' and 'effects' are measured, i.e., they are application tests which show performance against specifications demanded by statutory requirements or by the user or other parties and relate to a particular application and use of the product. As a result they convey to the potential client the kind of information he is most interested in, i.e., they show if the product in a specific application performs to a required standard which is fully intelligible to the potential user.

Additionally, the display of various charts, photographs and other visual aids are invaluable in enabling the innovator or others to present an unbiased view of the strengths and weaknesses of the project in terms of both technological and market factors. Fuller consideration of the role of the presentation is given when discussing the implementation of the exploitation strategies in Section 7.5, but it is clear that the proposal for the exploitation of the new product must display effectively all the options available to the client in order to ensure that his ultimate decision is made with full foreknowledge of the extent of any development work which may still be required of him, before the project attains commercial credibility.

7.5 PRESENTATION AND NEGOTIATION — THE FINAL GOAL!

In the preceding sections of this chapter we have explored the role of patents, prior-art, know-how, and confidentiality, in relation to the product, process, applications and marketing etc. The innovative chain of Fig. 3 seems almost remote and yet it holds together for us the logic of our progression in spite of the fact that we have been dealing with a complex of lateral and transverse events in which creativity has been seen in ideas, experiment, invention, design and a number of associated areas of activity, but at last we can see our goal in sight — the ultimate exploitation of the project, as depicted by Fig. 7, in Chapter 2.

In the eyes of the lone inventor-entrepreneur, a successful exploitation of the project is to be found in the sale or leasing of patent rights and know-how — on a basis of royalties and/or fees — to a business entrepreneur or manufacturer who will undertake the pre-production development of the product. This entails larger scale manufacture of the prototype product with which to sample the market prior to any major investment decision for full-scale manufacture.

The alternative of the inventor-entrepreneur securing venture capital and initiating the pre-production development stage himself is a massive challenge, but some will have sufficient confidence in their project to convince the investment houses and themselves that the project will be a business success as well as a technological personal triumph, by their readiness to make further investment from their own financial resources!

Where a project involves extensive 'processing' as in many chemical and manufacturing projects the scale of personal investment required from the inventor may be too great for him to contemplate, but he may have the option put to him by some institutions that he could surrender a portion of his patent property in the form of potential royalties from sales. In general the inventor will be reluctant to proceed down this course simply because he does not secure any cash return in the immediate future with which to sustain his innovation business, neither will he be able to predict at such an early stage the potential profitability of the project against which he could estimate and negotiate a fair and reasonable percentage basis for such an apportionment of royalties. What the lone inventor-entrepreneur requires is — as we say, 'something up-front' in any exploitation strategy if it is to serve his immediate needs. Optional terms are discussed below.

Assuming that the exploitation strategy is to capitalize on know-how and patent 'property' our objectives are now to secure a buyer for the 'property' and ensure a commitment to take the project from an established prototype demonstration stage of development to the point where a capital investment decision for the design and construction of a semi-technical or pre-full-scale production plant can be made.

Before any approach to the potential client is made, it is essential to review the optional financial terms open to consideration with respect to patents and know-how in any exploitation of a project as follows:

(1) Outright cash purchase of all know-how and exclusive manufacturing rights by the 'developer', i.e., no return on royalties on sales to the inventor.
(2) Acquisition of know-how and exclusive manufacturing rights by the 'developer' in return for an agreed royalty percentage on sales for the inventor.
(3) As (2) but combining a limited cash settlement with a reduced royalty percentage on sales for the inventor.

Of these three principal options the one which is likely to meet the lone inventor-entrepreneur's needs most adequately is in my view option (3). This arrangement provides him with 'working capital' while retaining his stake in the future profitability of his project under the management of his collaborator.

Securing these kind of terms depends on many factors, but the most important is the techno-commercial status of the project. If this is strong, the negotiating position should reflect it and option (1) should be avoided if possible, i.e. the inventor should aim to retain as much control as possible over his equity in the form of know-how and patent rights as long as economically feasible.

Sometimes a collaborator may favour an intermediate position in negotiations, i.e. that the inventor should offer an option rights agreement in return for some modest payment, which would allow a collaborator time to evaluate what is offered in the form of patent applications and know-how before deciding on the merit of acquiring manufacturing rights at some later stage.

Option rights agreements can be made on an exclusivity or non-exclusive basis. The difference here being that the exclusive agreement restricts the inventor from making any other agreements in whole or part with any third party. Under such a restraint the inventor ought to expect a larger option fee than he would receive under a non-exclusive option agreement and this should feature in any negotiation.

The corporate innovation team's approach to exploitation is conditioned by less critical cash-flow considerations than apply for the lone inventor-entrepreneur, but the range of options open to negotiation is still similar.

We turn now to the stage of preparing for the project presentation to interested parties for which our preceding thoughts on patents, know-how and confidentiality will have equipped us. Much of this information will have been compiled for the review at the end of the development phase (Fig. 6), but additional information will be required including the updating of the earlier data.

The key items to consider are:

(1) The data sheet.
(2) A market survey which will only be disclosed after Confidentiality Agreements have been formally signed.
(3) Samples of product including any prototypes available for display.
(4) Cost estimates of process and product must be presented as notional costings at this stage and the general basis on which they are made needs to be included, i.e., the plant design output and estimated utilization will feature. This will involve a notional arrangement and flow sheet for the development plant envisaged to take the project through to the full-scale production design sanction stage. This can be done without making any fundamental disclosure of 'how' the process is realized if Confidentiality Agreements have not been signed.

(5) A review of the technological status of the product with respect to competition is essential and a similar appraisal of the applications potential for the product.

With this information an exploitation strategy can be formulated around the negotiation options terms that we have outlined earlier.

In preparing the 'presentation' we need to ensure that it is specifically written to relate to the company or individual who is ready to consider seriously what the new venture project has to offer for their own business, perhaps by meeting specific market outlets in which we identify a potential business and technological fit.

We have already commented that the data sheet will display the key features of the product and/or process and this will feature in the introduction to the project along with an indication of the market envisaged. The display of samples of prototypes and the projection of photographic records of experimental evaluation and any statutory or user tests, utilizing where appropriate graphic aids, will provide valuable support for the information package and provide a cost projection of the process and product to demonstrate that cost/effectiveness and competitiveness can be anticipated from a commercial scale of production.

The presentation itself demands careful planning. The meeting place environment needs to be conducive to a relaxed introduction by means of a display of the prototypes or samples, all of which must be clearly labelled. The layout of the room must provide for wall charts, projection of photographic records, as well as the sample display.

I have found that providing the preparation has been thoroughly carried out, the reaction of those interested enough to discuss the presentation proposal is invariably very positive. Only in the larger company have I found isolated cases of individuals displaying a somewhat cavalier attitude to any attempt to introduce an innovative project into their own territory. We call it, 'the not invented here syndrome'!

The problem when encountering the latter attitude is that communications become distorted and individuals hear what they want to hear. The result can be catastrophic for the lone inventor as it can lead him into false expectations in terms of the levels of interest in his project and even involve him in tasks for the potential client for his project without a clear understanding being reached on the terms under which such work is to be carried out, with an inevitable disappointment to both partners of an option rights agreement.

Determining the potential client's level of interest and enthusiasm in a project is a vital preliminary to embarking on the exploitation phase negotiation.

8 The Practice of Innovation

8.1 INTRODUCTION

The wide spectrum of activities which make up new venture projects, targeted on the realization of manufactured products, has been considered from both a philosophical and practical standpoint in the preceding chapters, with particular emphasis on the role of key events and the respective phases that characterize the innovation process. In order to provide a comprehensive illustration of the implementation of the philosophy presented, I have turned finally to one specific case history, selected from a number of projects which I have undertaken in the capacity of a lone inventor-entrepreneur.

The selected 'case', will I hope, underwrite all the preceding philosophical points that have been made and also remind us that the 'practice of innovation' is a highly personal and practical endeavour in which, although the innovator is primarily attempting to confer practicality upon ideas in order to achieve a new product and/or process on which a manufacturing business could be established, within that objective and primarily in the role of inventor, he also seeks to express and fulfil his own creative urge in a way that so often mirrors that experienced by the visual artist, with whom he has much in common.

The analogy with the visual artist has been drawn in several places in the text and can be summarized in the way in which change, reiteration, mistakes, rejection, interaction and renewal, all occur in the search for completion and excellence. In a sense these aspects may appear to be remote from our accepted ideas of how research based innovation is pursued, but art and science do in fact often mirror each other in their interpretation of the natural world, as Waddington (1968) reveals to us in his study of the relationship between painting and the natural sciences — Behind Appearances.

The 'project case history' which I have selected for this 'rounding-off' stage, concerns the *innovation of a process for providing improved fire resistance to flexible open cell polyether-polyurethane foams, which find their application in furniture upholstery*. It is a project in which the mistakes that I made in some areas of its exploitation are evident and for this reason alone I feel that it will serve to complete the view of innovation that I have sought to present.

The idea for the project arose from two sources. The first of these originated with previous personal experience on a project concerned with introducing improved fire barrier properties into *rigid polystyrene foams* — a material apparently possessing little in common with a *flexible polyurethane foam*! The second source was in the identification of a wide area of public opinion, which gained authority and emphasis from the initiative of the UK Fire Services, in urging Government to prohibit the use of standard grades of polyurethane foams in upholstered furniture construction.

This preoccupation with safety and its focus on the hazard presented by polyether-polyurethane foams in particular was related not only to their ease of ignition, but to the highly toxic nature of the emissions that they give off during combustion. In addition they generate large volumes of dense black smoke that introduces a serious restriction on securing a means of escape in case of fire.

Providing a means of retarding the onset of ignition, reducing the rate of surface spread of flame and obtaining a corresponding reduction in the emission of toxic gases and smoke, presents a clearly defined objective for researchers and innovators in general. If this objective were to be realized it would be of considerable benefit to a large proportion of the population by making possible the introduction of 'safer' furniture that would lead to a progressive reduction in fire casualties as furniture was replaced over an extended period.

These arguments represented a strong motivation to undertake the development of fire barrier grades of flexible foams without delay. This is not to suggest that the motivation was simply or primarily a social concern, although it was a reassuring thought that the project had a humane ingredient in it. The point was that the market pressures were increasing for a new product which could be clearly defined in terms of 'effects requirements', as a definite research target. Evidently, if a new product could be realized to meet this target it would expect to command a very large market, providing it was supported by legislative requirement for its deployment and/or be available at acceptable cost to the users. It was now a case of — all systems go!

With these thoughts in mind I turned to the example I already possessed in the form of the fire barrier *polystyrene composite foam* described in Chapter 4. Although this foam was a rigid structure formed by bonding together pre-expanded foamed beads of polystyrene, the concept and technology appeared to me to be capable of adaption to the flexible open-cell polyurethane foams. An obvious advantage was that my colleagues and I had already developed the process for the *post-treatment of polystyrene foam rigid boards*. The process utilized inorganic layer minerals, and more specifically, exfoliated vermiculite, which was delaminated and classified for particle size in order to provide the inorganic fire barrier component in the foam matrix. The question of selecting the appropriate mineral additive was one in which particle size and geometry were of particular importance as it was essential that the particles formed a homogeneous structure based on film-forming properties.

Chapter 4, makes it clear that the creation of the *rigid polystyrene composite foam* depended on first modifying the structure of the polystyrene foam matrix so that interstitial porosity replaced the relatively impermeable structure of conventional polystyrene bead foams. Providing the foam with porosity, i.e. between the polystyrene partially expanded beads in this case, lay at the heart of the barrier concept which was to establish a reticular foam structure of the layer mineral within the organic foam matrix — effectively forming a highly continuous barrier to act as a containment of the organic material when subject to flame conditions. Fig. 12 provides clarification.

The method adopted for introducing the layer minerals into the interstitial pores of the foam matrix was by the simple step of making an aqueous dispersion slurry of the layer mineral, vermiculite in this case, and then immersing the polystyrene rigid boards in it to achieve impregnation.

It now becomes evident that achieving a similar post-impregnation treatment of *polyether-polyurethane foam* should prove to be a simpler proposition to that which was the case with the polystyrene foam, as the flexible grade of polyurethane foam matrix is of an open-cell structure, but differing in the fact that it is of a lattice form, instead of 'bonded beads'. Another advantage of the urethane foam with respect to the impregnation processes, of course, lay in its flexibility. This was to greatly facilitate impregnation, by compression cycling of the foam during processing.

What was not obvious at this stage was the extent to which the layer mineral could coat the surface of the lattice matrix of the *polyurethane foam* or compare with the excellent coating achieved on the relatively large surfaces encountered with the beads which formed the matrix of the *polystyrene foam*. Neither was it obvious that the inorganic coating in the *polyurethane matrix* would remain as an integral part of the foam structure when it was eventually submitted to service conditions, in which many rigorous deflections of the matrix would be involved. This product had to be sat on to ultimately prove itself!

The real challenge of innovation can now be seen to lie within those areas of uncertainty that also constitute the field in which novel solutions may be demanded. This is precisely the situation that the inventor can feed upon to secure invention and establish a patent basis with which his own business exploitation strategy must be established. At this stage it looked as if the idea that started the project could be realized.

It may be imagined that the next logical step would be to turn to an examination of any prior-art in the field of fire barrier upholstery, but such was my enthusiasm for evaluating the idea for a *fire barrier composite polyether-polyurethane flexible foam* as a result of these earlier considerations that I resolved to move to the experimental stage with minimum delay. The view I held was that it was *essential to know if the basic concept was feasible as early as possible* before committing resources to extended enquiries of the kind indicated, especially where these might involve making some form of disclosure on the direction of my own research.

Supplies of the polyether-polyurethane foam and the layer minerals were all that I needed to make a rapid laboratory-scale attempt at demonstrating the feasibility of the idea. Obtaining supplies of the layer mineral was soon to prove to be the first of a number of difficulties which had to be overcome, but it was also eventually instrumental in diverting my interest towards alternatives that strengthened the eventual claim of novelty for the product.

8.2 GETTING STARTED

Getting started, or taking the initial idea forward to implementation is always difficult. It demands a change in one's thinking and often a great deal of energy. Making the decision on what to do first is the breakwater point. It reminds me of a Czechoslovakian cartoon film I saw in which a lone figure starts his day by gazing through the window, finds little of interest, goes to the piano and lifts the lid, then drops it, turns round and stares unseeing at a blank wall then finally sits in his chair and moons away at an uninspiring world, gradually sinking lower and lower in both mood and position until he 'flows' out onto the floor. A case of hitting the bottom before he had done anything. In my experience it is 'in the doing' that we learn, ideas interact, mistakes yield information and invention is born. We had done our looking through the window and we now needed to do something about the ideas we had. Materials were needed, to be able to test out the basic concept. A decision had been made!

The vermiculite on which the *polystyrene* composite was developed was the obvious choice for the experimental work on the *polyurethane foams*, but this was a speciality product, involving a process of delamination and classification and was still at the development stage in ICI at the time. My efforts to purchase supplies was unsuccessful! A few litres of the vermiculite was all I needed for my simple research programme, but in view of my failure to secure samples of the delaminated and classified vermiculite, it was time to turn back to fundamentals again. I was after all learning the hard way, that the role of the innovator in the freelance inventor-entrepreneur capacity was a long way removed from that of the team innovator in the R & D organization, which I had enjoyed earlier. Getting started is harder than we imagine, especially for the man out on his own.

It was now a case of back to the drawing-board, or in this instance, to studying the data available on alternatives to the delaminated vermiculite.

Vermiculite is one of a group of materials, know as silicate layer minerals, all of which exhibit a high aspect ratio of planar dimensions to thickness. Hence their classification as 'layer minerals', which aptly describes their plate-like characteristics and which in the case of vermiculite has been shown to include excellent film-forming properties. The alternative material to vermiculite for the purpose in mind should evidently possess similar characteristics if it was to stand any prospect of achieving the composite structure desired.

Inspection of data on other layer silicates e.g. mica, bentonite, talc, etc, revealed that *bentonite* had characteristics which were similar in many respects to the classified delaminated vermiculite. Most importantly, the benonite platelet size is typically smaller than the vermiculite platelet, but basalt spacing is similar, suggesting that it should provide an acceptable alternative to vermiculite as an inorganic film-forming component of the composite. Also the smaller platelet size should be more suited to the task of encapsulating the lattice struts which form the 'open cell' structure of the polyurethane foam. It was hoped that the change to bentonite might produce 'films' which were tougher than those of vermiculite and therefore increase the chances of surviving the severe flexure conditions that the product would meet in service.

Another important feature of bentonite is that it occurs naturally in many quarters of the world and in a form that can be easily classified in terms of particle size, without resorting to complicated delamination procedures. This immediately points to it being a much cheaper feedstock than was the case with the delaminated vermiculite. A point that was quickly confirmed. Still another feature of bentonite was that the interlamellar bond strength compared favourably with that of vermiculite. It now seemed likely that bentonite held the prospect of achieving for polyurethane foam what vermiculite had done for the polystyrene foam. What was unpredictable, was the extent to which bentonite might meet the rigorous duties demanded by the final product. Only experiment would adequately answer that question.

I had reached the point where bentonite now appeared to be the most likely contender for the inorganic component of the modified 'urethane foam product' I was seeking. Obtaining supplies of bentonite proved to be a much better proposition than applied with the delaminated vermiculite. Steetly Berk Ltd were very receptive to my enquiries for supplies for undefined research purposes and made samples freely available to me. It was now a matter of securing supplies of the appropriate polyurethane foams.

Not wishing to expose my ideas to foam makers prematurely, I purchased samples locally, sufficient to meet my immediate feasibility programme.

I had got off the ground. The next step was to make the first experiment — this was exciting stuff!

8.3 EXPERIMENT AND FEASIBILITY

The result of the first attempt to impregnate the laboratory scale sample of 'urethane foam' with an aqueous slurry of bentonite in concentrations based on my earlier experience with polystyrene and vermiculite lived up to the initial expectations. Using repeated compression cycling of the foam, the slurry was quickly 'taken up' through the entire thickness of the sample which was 50 mm. Drying the impregnated foam — soon presented itself as the difficult and prolonged feature of the process, but it was soon evident that the way to proceed was with convection hot air circulation. The resulting dried sample displayed some rigidity which was quickly removed by post-compression cycling.

It was time for the fire performance assessment. A number of British Standards, provided me with an indication of the type of test which was being used to assess foam mattresses and this took the form of ignition by small cribs of wood which were placed at strategic positions on the sample of mattress, but the test was inappropriate in this case as my samples did not meet the test sizes specified. Another test was specifically designed for upholstery cushions, known as the cigarette test. Other tests used newspaper as an ignition source and some a simple match ignition. My need was for a simple and quick 'indicative test' to enable me to form an appreciation of any 'effect', at this preliminary stage. A conventional propane gas torch test was eventually selected for this purpose as it was severe enough to show up only distinctly positive fire barrier effects.

The result was surprising. The flame emission was low and quickly ceased where the organic material had been burned away, but smoke emissions were minimal and the inorganic reticular layer mineral structure within the foam matrix retained its integrity. In particular no significant flame spread was evident. The foam did not drip or collapse nor change significantly in its dimensions. As a 'control', an untreated sample of the same grade of polyurethane foam was ignited with the propane torch and the result was relatively devastating. Rapid ignition and flame spread, plus dense black smoke followed by eventual total collapse of the sample.

It was a 'comparative' result that could certainly form the basis for a claim that a post-treated polyurethane foam employing a layer mineral, i.e. bentonite, as the 'barrier material' achieved a marked increase in resistance to flame ignition and significantly reduced smoke levels while retaining structural integrity. The desired fire retardancy effect had been demonstrated and the process for providing the 'effect' had been shown to be feasible.

Much more was needed before imagining that a product was within reach, but a start had been made and it was now essential to take stock of the situation.

Repeating the experiment with laboratory scale samples is an essential step and this was carried out with similar results. The next move was to consider the question of scale-up to sample sizes which conformed to some of the BSS Fire Tests referred to earlier. These required the introduction of some degree of quality control in their preparation. Procurement of samples from the foam makers was now essential in order to secure reliable product specifications of the samples to be used in the tests, but before making this approach to them it was also necessary to consider filing a patent application on the 'new product' and process which had now been demonstrated.

8.4 PATENTING STRATEGY

In my case I enjoyed the services of a Patent Agent who was extremely helpful on a number of aspects of the project. In addition to interpreting my outline

Patent Claim in the form of patent applications he was also instrumental in challenging me to produce specific data with which to add credibility to my claims. Other benefits I received were his participation in the ultimate Options Rights Agreement negotiations I had with potential clients for the patent property, but the key benefit was that he was able to guide the implementation of an effective patent strategy, which had to consider which countries must be included in any application and balance the desirability of patenting in whole or part the option of retaining 'confidentiality' over key areas of know-how.

Obviously we needed to start with the country of origin, but in my case it was decided to extend this to the EEC and eventually to the USA. The point here was that methods of increasing the fire resistance of upholstered furniture is of world-wide concern and involves manufacture on a similar scale. Hence world-wide patents could be essential in order to ensure effective exploitation of any patent property. Making the claims was now implemented.

The basis of the patent claim had to rest on the experimental data obtained and took the form of a brief description of the product, its analysis and its performance as characterized by specific tests and evaluation procedures. Also included was the description of the process in sufficient detail for others to replicate the product and its effects.

The particular novel elements of the product and process and the 'effects' or 'properties' obtained were highlighted to provide the Patent Agent with a basic picture of the project on which he could expand the basis of any claim. This is an extremely vital part of the project as its commercial value for the innovator depends so much on showing that the novelty does exist and that all the appropriate implications of this are reflected in the patent claim.

In characterizing the product it was essential in this case to specify the mass loading ration of layer mineral to foam which realized the barrier effects observed in the test programme. This proved eventually to be of critical importance to establishing the novelty of the 'product'.

What happened at the patent search stage was that an earlier patent relating to the use of bentonite impregnation of flexible foams was found. This had the aim of improving the wetting-out of synthetic sponges, i.e. to improve hygroscopic properties. On examination it turned out that the percentage loading of layer mineral stipulated in the patent was extremely small, i.e., a few percent by mass. This proved to be of great significance to the prospect of gaining a patent for the new product as the experiment with bentonite and polyurethane foam indicated that a loading ratio as low as advocated in the 'sponge' patent would not provide the foam with a sufficiently strong barrier film to attain the fire barrier properties required, nor would it result in a strong 'char' structure which resisted collapse during flame impingement. The higher dry mass loadings specified for the barrier composite foam were now evidently a critical basis for the patent claim.

8.5 WEIGHING UP THE OPPOSITION

With a Patent Application filed and confidential know-how secured, it was appropriate to approach the foam makers to explore the supply of sample materials, to see if a collaborative evaluation was possible and to explore any other avenues that might prepare the ground for eventual project exploitation with them.

It was important in selecting which foam maker to approach, to ensure that they commanded a good proportion of the UK furniture upholstery market with their product. This proved to be Harrison Jones Ltd, in the UK. The result was that on making contact, sample supplies were secured, non-confidential information exchanged and an offer was made by the foam makers to carry out tests on the new product, generally in accordance with the BSS upholstery specifications etc. This meant some degree of disclosure, but providing it was not in excess of the Patent Application claim it was an acceptable arrangement to enter into. Of course the product formulation and details of processing were kept confidential at this stage even although a Patent Application had been filed, as it would not be published until the statutory eighteen months after filing had elapsed.

The level of co-operation secured was excellent and this was the result of establishing contact with the director responsible for new developments within the technological area. An additional benefit was that the director displayed every encouragement and goodwill towards my efforts. Making the right contact is of considerable importance to the inventor-entrepreneur!

The foam makers 'return', was to gain information on the rate of progress on my project, which might be of importance to his own company's development strategy, while I gained technological and market data, so far as confidentiality constraints allowed. This included information on the general level of the 'state of the art' in the development of alternative solutions to the problems of foam fire hazards. It was a pleasing arrangement I believe to both sides and especially invaluable to an inventor-entrepreneur with limited resources.

I was subsequently to learn that although a variety of materials were being researched as candidates for post-impregnation of flexible foams by competitors aiming at this particular market, the layer minerals, with their novel film-forming characteristics, appeared to have been ignored. Obviously it is important to measure the performance of any new product against that which can be achieved by competitive products as well as attempting to comply with any statutory or other form of regulation or specification requirements. With this in mind it was a further step forward when the foam makers offered me the opportunity to submit samples to them for comparative testing against the performance of competitive products on the basis of our joint observation of the tests. Such arrangements are vital to the advancement of any new product, but for a lone inventor-entrepreneur they represent a most invaluable aid to his research programme. The willingness to co-operate in this way is as we have seen inevitably tied to the business self-interest of the foam makers but this is not to belittle that sense of goodwill displayed towards the man with another approach to a new product.

The picture I now had of the project's potential for advancement had improved rapidly by these latter events which confirmed that the way I had approached it was closely in line with the thinking of others, which in one sense is encouraging, but in another, a warning that achieving novelty and with it an exclusive exploitation position might yet prove difficult. The strength of the project's exploitation potential, however, clearly resided in the selection of bentonite as the novel impregnant material.

One further enquiry was needed in 'weighing up the opposition'. It was to ascertain if the foam formulation could be modified by the introduction of flame retardants at the 'nozzle' end of the existing process line, i.e., could the need for post-treatment be circumvented by an on-line process modification? At the time it appeared that while it had not yet provided a satisfactory alternative to the post-treatment systems, work by other researchers was continuing in this field.

It is evident from this case history that a considerable amount of the new venture project is spent on exploring what other workers are up to. This is vital in order to gauge the level of commitment and resources which should be allocated to the project. In some cases this kind of market intelligence can lead to the abandonment of a project, but in this case what finally emerged was that a post-treatment system appeared, at this point in time, to be the most likely means of realizing the product.

This conclusion now supported the direction taken with the project from the start and it is one particularly suited to the resources of the lone inventor as the principal feedstock material was already available in the final product form, i.e., as flexible foam slab, 'only' demanding the introduction of a fire barrier treatment system, which of course is very much an understatement. Fortunately the range of research involved can be contemplated without recourse to the kind of high capital and revenue expenditure frequently demanded of many industrial research projects. Keeping an eye on the budget suddenly takes on a new significance when the project and the funding depends on one's own resources!

Significantly the post-treatment approach lay within my own knowledge and technical capacity for implementation, whereas the research demanded to effect an on-line chemistry solution to the fire retardancy problem lay outside my own capability. For the lone innovator it pays to focus on projects for which he possesses a background of know-how and scientific or technological knowledge. At least it must lie within his capacity to acquire such knowledge if the project emerges in 'new territory'.

8.6 SCALE-UP AND APPLICATIONS

As a result of establishing contact with the foam makers it immediately became evident that while the initial samples and tests demonstrated that a new composite foam product might be realized the form of the product required some review in relation to scale-up.

The applications envisaged from the start, were to produce the foam composite to a specification that would be capable of satisfying the manufacturing and use requirements of foam cushions as used in upholstery, including foam inserts for special shapes etc. In addition larger scale forms appropriate to applications as foam mattresses was recognized as another viable target.

In relating the experimental results to these application targets it was appreciated that the foam thickness required would have to accommodate a range of anything from 1–10 cm. The experiments indicated that when impregnating the samples at the thicker end of the range it was inevitable that the concentration of layer minerals was largely confined to a region of approximately 1 cm deep at each exposed face, leaving the core untreated. This was due to the filtration effect of the foam cells which extracted the mineral from the slurry as it penetrated the surface regions. It followed that a sample of approximately 2 cm thick could be satisfactorily impregnated, i.e. irrigated from two opposite faces to achieve full impregnation using the simple compression cycling process already developed. Earlier laboratory scale samples of 5 cm thick had been fully impregnated, but this result was achieved with foams of lower density than in the scale-up series of experiments.

These observations were instrumental in providing a series of four optional product forms for the applications concerned. In terms of the Innovation Project Model this represented the contribution that is made by 'design', to the realization of the product prototypes.

The four optional product forms were:

Design Version (1)

The simplest and basic product form was evidently a thin sheet of foam with a thickness of approximately 2 cm (Fig. 23). A particular merit of this form can be recognized in the facility it affords for continuous line processing, in comparison to the batch processing demanded by specific bulky forms of the product.

Design Version (2)

In order to achieve thicker foam sections, i.e., >2 cm, the design employed consisted of bonding thin sheets of the composite foam, to all faces and edges of an untreated foam core (Fig. 24(a)), i.e., producing an encapsulation design of upholstery cushion.

Design Version (3)

The construction of a form in which the entire matrix of the structure consisted of treated foam was attainable by laminating several 'sheets' of the composite

Figure 23 Flexible polyether-polyurethane/layer mineral composite foam — sheet form

foam together to achieve the desired thickness. In this case the laminates were typically 2.5 cm thick (Fig. 24(b)).

Design Version (4)

A distinctive option, representing the initial concept used in the feasibility phase of the programme, was also possible for those forms > 2 cm. In this case instead of relying on the use of laminates, slab of any thickness up to say 10 cm, was treated by surface impregnating to the established depth of 1 cm, leaving the core material untreated. The advantage of this form is that it avoids the tedious and costly fabrication methods of the multi-laminate structure entirely. (Fig. 24(c)).

In the latter example the treated zone thickness was reduced to 1 cm as the impregnate could now only be introduced to this depth at this stage of development. The fact that the core was untreated was similar to the encapsulation design version. Both versions use the composite foam layer or zone, as a protective barrier for the underlying untreated foam.

By a combination of experimental observations and a study of the processing limitations, the design versions described now emerged as candidates for applications research, i.e., prototype sample product was now required, to implement the design concepts and for evaluation against specific tests demanded for the selected applications. It is evident that the applications phase and the development phase are closely interwoven in this example, but each can be seen as distinctive contributions to the innovation process.

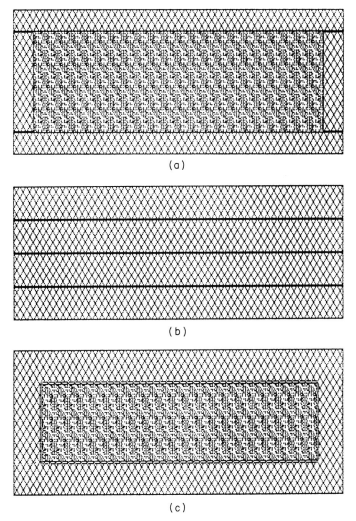

Figure 24 Barrier foam upholstery design options. (a) Encapsulated upholstery form. (b) Stacked laminate upholstery form. (c) Zone impregnated upholstery form

The project provides a clear example of the transition from the preoccupation with the fundamental concept, interpreted by the 'matrix' of the product, to the macro-composite ideas of the four design versions of the product which are application related.

8.7 PROTOTYPE EVALUATION

Evaluation of the various product designs was now to be based on matching the particular product form with a specific application and consequently with a

(a)

(b)

163

(c)

(d)

Figure 25 Flexible polyurethane/layer mineral zone impregnated composite foam upholstery sample. Fire test stages. (a) At ignition. (b) After 5 minutes. (c) After 7 minutes. (d) After 11 minutes

(a)

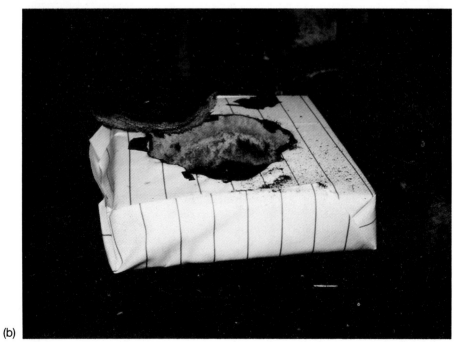
(b)

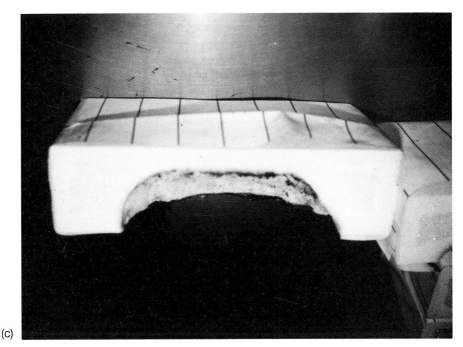

(c)

Figure 26 Flexible polyether-polyurethane/layer mineral zone impregnated composite foam. Cross-section of flame-tested upholstery sample. (a) After completion of fire test. (b) With 'char' crust removed. (c) Cut cross-section with 'char' crust removed

particular test requirement, generally in accordance with purchase specifications or manufacturing standards.

In common with many other projects at this stage it was considered that the first step was to be a simplification of the 'approved tests', i.e. in the form of 'indicative tests', on the basis that speed of testing, and economy of materials were important at this stage. Also that the tests could be conducted 'in-house' and lead to rapid changes in specification if required. The invaluable support of the foam makers now becomes evident. They had already developed the type of screening test referred to, if they had not done so, it would have placed a considerable work-load on my own programme, but now it was possible to start testing as soon as prototypes were available.

The first test series was carried out with the zone impregnated version of the product (Design version (4) (Fig. 24(c)). The first test record is represented by Fig. 25(a)–(d). The end result, as shown in Fig. 26(a)–(c), was that while the internal untreated zone melted, the treated zone formed a stable char structure which contained the flame and to a marked extent, the smoke during the test. The char section was lifted off the sample to expose the cavity formed within the foam due to the retarded combustion of the untreated zone. The strength of the char is evident from the fact that it could be handled in the way described.

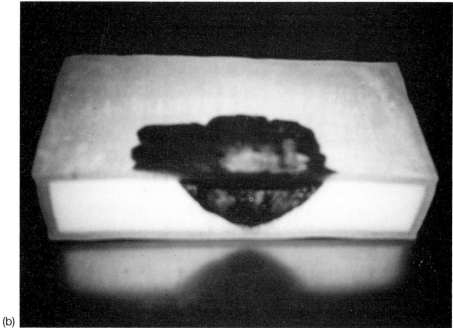

Figure 27 Flexible polyether-polyurethane/layer mineral encapsulation design version foam — cross-section of flame-tested upholstery sample. (a) After completion of test. (b) Cut cross-section showing integral 'char'

The next test series was with the encapsulated design version (2), of the product (Fig. 24(a)). Similar results to those observed with the zone treated product were realized, but in this case the integrity of the composite laminate layer was even better. The fire test sequence was similar to that of Fig. 25(a)–(d).

The sample was again cut through to reveal the relative performance of treated material compared with untreated material, as shown in Fig. 27(a) and (b).

The results of tests on both prototype designs were highly encouraging. They were in fact a replication of the effects seen with initial laboratory samples, but with the added value of demonstrating the scale-factor, i.e. that the spread of ignition damage in the case of the surface type fire test, was adequately contained. It also showed that although a much larger mass of material was exposed to flame, it did not generate excessive smoke.

In a different type of test, the laminated form of prototype (Design version (3), Fig. 24(b)), was cut into a number of smaller sections, which were then stacked to form an igloo configuration, ignition was achieved by placing the ignition source inside the 'igloo'. Although the result was the eventual collapse of the foam sections it was seen to be attributable to the fact that the stacking arrangement used was in this instance unstable, with the result that sections had fallen into the ignition source as the dynamic conditions of the test caused the stack to displace, i.e. failure was not due to a prior collapse of the foam matrix. The foam matrix retained its structural integrity.

By now it was reasonable to assume that the product might be capable of development to a commercially acceptable specification and be appropriate for a range of applications as indicated earlier, but more was demanded of the product than its fire barrier and smoke reduction performance. This was that the layer mineral additive would not migrate in normal use conditions and that the composite foam would also satisfy the same tests criteria for acceptability in terms of thermal stress recovery as the untreated foams did.

It did not take long to find that on these last two counts, more research was needed. Mass loss was too high when subjected to cyclical compression pounding and thermal stress relaxation was too low when evaluated with compression/thermal loading and release methods. Secondary invention was needed!

The prototypes were at the halfway stage. Ideas for resolving the problems took time to materialize, but eventually a method of anchoring the layer mineral in the matrix, i.e. bonding, was identified. This potential solution also offered a way of overcoming the thermal stress relief problem. Two for the price of one! I had entered the heart of the development phase.

It was time to look around for more resources.

8.8 EXPLOITATION

At this point it became necessary to consider if the next step was to raise capital for a major scale-up in the research and development of the product with a view to undertaking its manufacture myself on a small pilot scale process plant, or alternatively to seek out a client who would purchase option rights and know-how with a view to completing the development phase himself and then go on to design a manufacturing plant, perhaps with my support in a consultancy capacity on both activities.

It so happened at this time that a client was forthcoming who indicated an interest in the project and wished to explore the salient features of its potential in relation to his plans for expanding into businesses based on 'effects' type products. His particular interest was in a secondary version of the product, i.e., a rigid form of the urethane based composite foam, suitable for use in building applications, but the flexible form was also to be considered. This now represented an important breakthrough in terms of exploitation of the project and the future for it now looked very promising, but as events were to show, many disappointments were yet to present themselves as the exploitation phase was pursued.

Before proceeding any further along the lines indicated by the general exploitation strategy outlined in Chapter 7, a Confidentiality Agreement was drawn up and agreed to by both parties. A detailed presentation of the project could now be made to the client in which the chief features of product and process were presented, warts and all. Samples of product were displayed and test data tabled. A broad outline of the type and scale of market opportunity was introduced and an order-of-cost estimate of the product for various scales of production was given with comparative figures of potentially competitive products.

It was made clear during the presentation that the project status, was that while the process and product feasibility had been demonstrated and supported with a limited amount of development effort, a major development phase programme was demanded to bring the project to commercial realization as a new business venture. As a first priority, it would be essential to resolve the outstanding technological problems referred to earlier.

The purpose of this emphasis at this introductory stage was to ensure that the client would have no doubt about what was entailed for him in adopting the project, i.e., he would be committing his own resources to a research and development programme, in order to advance the product to a stage where he could make his own assessment of its potential, with a view to investing further resources in it as a new business venture. This meant a commitment to the development of a small-scale manufacturing plant for the production of prototype products.

The client indicated that the project fitted their portfolio of new venture projects and their overall strategy. A follow-up meeting to review the basis for acquisition was planned. So far so good!

Before getting into the next meeting it was essential to review with the Patent Agent, the terms which should be negotiated with the client, and to enlist his participation in the next meeting, when the various options were to be discussed.

Clearly where a new venture is still at the patent application stage, the outright assignment of it to the client is not always feasible, as its commercial value may be difficult to assess. In these cases the assignment of option rights is a more reasonable arrangement from the standpoint of both parties. It leaves the assessment of the eventual valuation of the project know-how and patent rights to be debated at a later date when the project has reached the full-scale investment decision point.

In negotiations at the next meeting I was to lose some of the initiative, as the idea of option rights was seized upon by the client who interpreted it as an opportunity to secure an option period on financial terms which in retrospect were loaded in his favour.

The inventor-entrepreneur really needs 'something up front' to use the American expression, i.e. I needed a cash return to fund my work in whole or part. The financial terms put to me were that a sum equal to the cost of extending the patent applications to include the USA and the EEC, should be paid by the client as the fee for acquiring exclusive option rights. The client evidently appreciating that the wider market presented by these countries was essential to his own exploitation strategy.

The argument was that if the client did not continue with the project, I would have the advantage of still holding applications which were applicable to a wide range of countries and therefore of greater commercial value to me. Such a settlement meant that I had secured nothing in the form of cash 'up front', except that which had to be immediately expended on extending the patent applications to areas which served the long-term interest of the client. What I failed to appreciate was that the cost of servicing the USA application in particular is a lengthy and costly business and I was now liable to this cost, which would arise after the option term had expired, i.e., if the option rights were foreclosed. In other words I would have to put more financial investment into the application if I was to realize any concrete return on my option rights terms. It was a poor show for a novice in the negotiation stakes!

That such tight conditions were pressed by the client ought to be a warning to any would-be entrepreneur. In my case I accepted them in and the fond hope that all was going well, but I had not foreseen the implications of the terms agreed, quickly enough.

This now brought me to another area of considerable importance for the entrepreneur. It concerns the negotiation of consultancy agreements. In this project I was specifically asked to agree to act as a consultant to the client when the option rights agreement was finalized. Indeed, it was made a condition of negotiating the option agreement, but in spite of this insistence, the client also wished to defer any negotiation of consultancy terms until the option agreement was finalized. Once again I agreed to trust in a future situation instead of hardening it up in a formal document. Another mistake!

By now the reader will begin to wonder just what kind of an individual could make such a mess of the negotiation, or he may feel that I was being taken advantage of by a more powerful adversary, as opposed to a partner.

Figure 28 Rigid board version of polyether-polyurethane composite foam product — with surface coating

Worse was to follow; with the Option Rights Agreement signed, the client failed to put into the project the kind of resources he had intimated and turned to me to carry out all of the sample preparation needed for his evaluation purposes. I had still not received the formal offer of consultancy terms but, as I was apparently a glutton for punishment I undertook to carry out the work requested. On receiving details of the sample and test requirements it was obvious that it was necessary to undertake a development programme, involving product specification, method of manufacture and screening tests, as the samples required were to meet a new application for which samples had not been previously made.

My request that the work should be quickly related to a consultancy agreement incorporating terms of payment as intimated previously was made while I pressed on with what I saw to be a collaborative enterprise. Regrettably the response was not what I expected. It took many months before I obtained settlement of my invoice.

This story serves to illustrate that option rights agreements do not incur any liability on the part of the inventor to carry out any development or sample preparation and emphasizes the need for formal contracts to be negotiated before any support work is undertaken by the inventor-entrepreneur on behalf of the client.

Eventually the client in question did not renew his option, but I had the satisfaction of knowing that the samples he had requested from me had in fact passed the particular statutory test in which he was interested. My crash programme to develop this specialized version of the product, i.e., in the form of a rigid thermal insulation fire retarded board (Fig. 28), instead of the flexible form on which my work was primarily targeted, had at least advanced the product's technical options one stage further. The product is still live and I am a little wiser, or so I would like to think.

It is evident from this final project example that my own limitations have been revealed, but as I indicated at the start of this book, this has been done in the hope that it will help to inform others embarking on the entrepreneurial adventure of the problems that can be encountered. In this particular case, the importance of getting terms formalized early in the life of a project and ensuring that the implications of the terms negotiated are fully appreciated by both parties, before concluding any agreement, was of obvious yet paramount importance. Also when dealing with the larger companies, where project responsibility is often transferred internally from one group or department to another, at some intermediate stage during the option period, it is essential to ensure that the new project leader fully appreciates the status of the project he is evaluating. Failure to do so was I believe the cause of the problems I encountered. It is not enough to depend on the internal communications of the client's company as a means of ensuring continuity of the project in these circumstances.

In drawing this story to its close, the important point is that even the dissappointing experiences are assimilated as learning and in the project in question it led me back to the point where the Innovation Project Model always begins, i.e. to idea generation, experiment and invention. As a result I started to develop secondary ideas relating to the attainment of improved forms of polyurethane fire barrier flexible foams which hopefully may extend the technology and enhance opportunities for product exploitation. At the time of writing, the saga continues!

In conclusion, putting the product innovation experience into perspective, I believe it to be one of the most invigorating outlets that any creative individual, possessing the benefit of technological training, can in my view hope to find. The contribution I have sought to make to the subject is summarized by the innovation project model concept, where logicality determines the broad lateral progression of the project as depicted by the innovation chain in which we find transverse displacements, or 'shifts' in direction which are generally unpremeditated and hold the core of the creative input from which invention at times results. In retrospect the concept of product innovation which I have explored in this book might be summarized as embracing both a knowledge base and a creative base.

References

Birchall, J.D. (1983a) Cement in the context of new materials for an energy expansive future. *Philosophical Transactions, Royal Society, London*, A310, pp 31–42.

Birchall, J.D. (1983b) On the relative importance of relevance and irrelevance. John D. Rose Memorial Lecture: Chemistry and Industry.

Birchall, J.D., Bradbury, J.A.A. and Dinwoodie, J. (1985) *Handbook of Composites*, Ch. IV. Strong fibres. W. Watts & B.V. Perov Royal Society and Soviet Academy of Science. Elsevier Science Publishers BV.

Bradbury, J.A.A. (1977) Alumina fibre monolith support system for catalytic convertors. *Proc. Int. Sym — Automotive Technol., and Automation, Wolfsburg*. Vol 1, Paper 3.

Bradbury, J.A.A., Rolands, B. and Tipping, J.W. (1981) Organic inorganic fire barrier structural foam composite. UK Patent No. 8118638 — assigned ICI.

Carson, J.W. (1974) 'Three-dimensional representation of company business and investigational activities', *R & D Management*, Vol. 5, No. 1.

Department of Industry (1982) *Support for Innovation*.

Drucker, P.F. (1985) *Innovation and Entrepreneurship*. Heinemann.

Gardiner, P. and Rothwell, R. (1985) *Innovation — A Study of the Problems and Benefits of Product Innovation*. The Design Council.

HM Patent Office (1982) *Applying for a Patent*.

HM Patent Office (1985) *Protection of Industrial Designs*.

Parker, R.C. (1980) *Guidelines to Product Innovation*. British Institute of Management.

Parker, R.C. and Sabberwal, A.J.P. (1971) 'Controlling R & D networks', *R & D Management*, Vol 1, No. 3, pp 147–153.

Waddington, C.H. (1968) *Behind Appearances*. Edinburgh University Press.

Index

Added value 23, 27, 53, 102
Aesthetic appearance 36, 38, 125
Aircraft fire barrier seat 56
Alumina fibre
  base product 102
  filter use 128
  insulation 95
  mats 76
  products Fig. 15, 101, 107
  properties 101, 103, 118
  stacked block Fig. 14, 107
Alumina silicate 113
Analogy 8, 9, 13, 98, 99, 122, 150
Analysis
  cost 92, 136
  data 156
  event sequence 19
  experiment 75
  innovation 1, 2, 13, 14, 63, 70, 73, 74
  invention 94
  performance 125
  problem 129
  project history 75
  properties/effects 25
  quality control 37
  stress 48
Applications 5, 12, 14, 15, 65, 158
  awareness 27
  experiments 28
  ideas 30, 74
  innovation 5, 15, 97
  inventions 31
  market development 111
  patents 33, 120
  Phase 2 15, 17
  phase events 28
  phase model Fig. 5, 29
  phase review 16, 34, 121
  product 2, 11, 16
  product cost estimate 33
  product/effect evaluation 16, 32
  patents 33, 120
  product costs 33
  product definition 32, 33
  prototype 65
  quality control 32
  reports 17
  research 13, 15, 16, 29, 37, 65, 66, 79, 97
  secondary 16
  specialist 5, 38, 51, 99
  specifications 32
  strategy 15, 26
  tests 146
  targets 28
Areas of uncertainty 152
Artist analogy
  art and science 150
  concept 8
  inventor 150
  redefinition 8
  reiteration 9, 13, 122
Awareness 46

Barrier foam upholstery designs Fig. 24, 161
Base product 15, 16
  consolidation 28
  definition 25
  starting material 29
Batch processing 126
Batch production 36

Bentonite 154
Birchall, J.D. 65, 101, 107, 134
Bisphenol S 108
Bradbury, F.R. 102
Bradbury, J.A.A. 80, 114
Brain-storm session 7, 65
Break-even point 42
British Standards Association 51, 100, 129
Building panel 36
Building Research Laboratory 51
Business
  capital investment 17
  decision point 14
  interests 139
  management 17
  objective 138
  opportunities 4–7
  organization 7
  proposition 14
  strategy 23
  study 16
  support 55
  techno-commercial structure 31
  ventures 1, 3, 13, 14
Business Group 52

Capital investment 13, 14, 17, 38, 41
Capital estimate 38
Career ambitions 55
Carson, J.W. 66
Case histories 1, 2, 3, 10
Case history
  alumina fibre 94, 95, 101–106
  bisphenol S 108–111
  catalytic converter supports 112–119, 135
  diesel exhaust filter 128–133
  fire barrier composite rigid foam 80–88
  fire retarded polyurethane foam 150–171
  high strength cements 134, 135
  inorganic rigid foam 133
  tall stack design 48, 125
  wall cavity seals 75
Cash flow 41
Catalytic converters 113
ceiling tile 80
Ceramics Research Association 116
Change 6–9, 14, 6
Client 149
Code of Practice 50, 129
Collaboration
  market 30, 31
  user 37
Collaborative evaluation 157

Commodity chemical 110
Competition 37, 139
Composite foam test sample Fig. 12, 85
Composite materials 108
Concept 2, 6, 9, 15, 48, 70
Conceptual fit 48
Conditioned response 48
Confidentiality 24, 26, 121, 139, 140, 141
Conflict 8, 48
Consolidation 15, 28, 97
Consultancy 10, 18, 44, 169
Contracts 44
Control limits 119
Cooperation, research/marketing 66
Cost 16, 28, 59, 94, 111, 120, 134, 135
  plant, full scale 17, 38
  plant, semi-technical 17, 38
  process 17
  product 17, 40
  savings 59
  notional 26
Cost-effective product 27, 28
Creative 1, 2, 4–9, 14, 54, 61, 46
Creativity
  experiment 61
  limits 54
  student development 63

Data 15, 16
Data base 59
Data selection 42
Data sheets 17, 24, 25, 145
Decision 14–17
Definitions 2, 4, 6, 8, 9
Department of Industry 59
Department of Trade 51
Design 3, 6, 9, 14, 16, 105, 125, 126
Design Centre 10
Design Council 10
Design
  development 16
  events 16
  notional plant 41
  objective 9
  options 159, 160
  phase 15
  plant, full scale 13, 17
  plant, semi-technical 16
  process 1, 16
  product 16
  prototype 16
  sanction 13
  target 9
  team 115

Developed sensitivity 47
Development 2, 5, 6, 10, 14–17
　analysis 8
　contracts 16, 18
　creativity 4, 8
　events 11
　innovation 10, 15
　invention 9
　new materials 14
　option rights 17
　product 5, 6, 17
　proposals 33, 34
　prototype 17
　strategy 16
Development and design 123
　Phase 3 14–17, 34
　phase events 34
　phase model Fig. 6, 35
Development phase review 16, 17, 41
Development programme 170
Development proposals (laboratory scale) 33, 34
Diesel exhaust filter Figs. 20, 21 and 22, 130–133
Discipline 5, 48
　innovator 48
Disciplined 4
Disclosure 141
Drucker, P.F. 3, 10
Dynamic
　coupling 46
　process 6–9, 13

Economic
　aims 112
　growth 9
Economics
　process 16
　production 109
　research 16
EEC 144
　legislation (patenting) 26
Effects 15, 110
　product basis 108
Element concept 130
Employment 60
　pattern change 60
End forms Fig. 15, 107
End product 1, 8
Engineers 37, 63
Entrepreneur 4, 10, 16, 17
Environment 5, 7, 8
　innovator 46
　team 47

Epoxy 108
Estimates 41
Evaluation 15–17, 24
Evaluation event 24
Event
　No. 1: Idea 21, 74
　No. 2: Experiment 22, 87
　No. 3: Invention 23, 79
　No. 4: Evaluation 24, 88–92
　No. 5: Basic product definition 25, 88
　No. 6: Patenting 25, 26, 93, 120
　No. 7: Preliminary product costing 26, 94
　No. 8: Feasibility phase review and applications strategy 26–28, 95
　No. 9: Base product consolidation 28
　No. 10: Applications experiments 28–30, 87
　No. 11: Applications ideas 30, 31
　No. 12: Market research 31, 92
　No. 13: Applications invention 31, 32
　No. 14: Applications products evaluation 32
　No. 15: Application products definition 32
　No. 16: Application patents 33, 120
　No. 17: Application products costing 33, 94
　No. 18: Development proposal laboratory scale 33
　No. 19: Applications phase review and development proposals 34, 121
　No. 20: Prototype design 34–36
　No. 21: Laboratory process unit design 37
　No. 22: Applications research/prototype 37
　No. 23: Market research/prototype 38, 111
　No. 24: Semi-technical plant costings 38
　No. 25: Full scale plant costings 38
　No. 26: Prototype evaluation 39, 40
　No. 27: Product costings 40
　No. 28: Project data summary 41
　No. 29: Development phase review 41, 42
　No. 30: Exploitation strategy 42
　No. 31: Data selection 42
　No. 32: Patent review 43
　No. 33: Option Rights 44
　No. 34: Consultancy and contracts 4
　No. 35: Exploitation proposal 44
　No. 36: Presentation and negotiation 45
　No. 37: Project adoption 45

Events 1, 6, 8, 11, 17, 20
   adjacent 12
   critical/key 11, 13, 14, 18, 20
   definition 14
   design 16, 19
   developed sensitivity 47
   feedback 21
   innovation chain 14
   innovative 6
   interactivity 11, 13, 21
   inventive 14
   inventory 20
   key 19, 1
   paired 13, 72, 73
   pattern 9, 11–13, 121
   permutations 12
   scale 11
   sequence 8, 12, 19
   single 13, 70–72
   triple 13, 73, 74
   unit operations 14, 19
Evolution 1, 15
Exhaust systems 113
Experience 1–5
Experiment 2, 6, 7, 13, 15, 87, 154
Exploitation 1, 4, 13–16, 167
Exploitation
   business 1, 7, 15
   goal 9
   idea 4, 5
   invention 5, 16
   options 139
   product 2, 9, 11
   proposal 44, 96, 145
   strategy 15, 17, 20, 42, 138
   technological 5
Exploitation phase events 42
Exploitation phase model Fig. 7, 43

False starts 13
Feasibility 14, 15, 17, 21, 27, 70
   product 15
   project 7
Feasibility phase events 21
Feasibility phase model Fig. 4, 21
Feasibility phase review 26, 95
Federal Aviation Authority 56
Feedstock 7, 15, 16, 21
Fibre mat Fig. 13, 103
Field of use 25
Financial
   grants 58
   option terms 147
   resources 66

   risk 53
   support 62
Fire barrier building panel 36
Fire barrier composite foam 152
Fire barrier structural foam 80
Fire performance assessment 155
Fire testing 95
Functionality 11
Furniture Research Laboratory 51
Furniture upholstery 151

Gardiner, P. 10, 11, 13, 64, 65, 101, 118
Getting started 153
Government research laboratories 50

High speed cure resin 109, 110

ICI plc 52
Idea 1, 4, 6, 8, 9, 21
Idea and invention (distinction) 77
Idea
   applications 22, 30
   brain-storming 7
   change 9
   corporate search 6
   environment 7
   exchange 59
   expansion 8
   experiment 6, 22
   exploitation 4, 5
   field 5
   generation/emerge 1, 4, 6, 7, 14
   implementation 3, 6, 13, 14
   innovation 2, 6, 74
   innovator 7
   interactions 13
   intuitive 14
   invention 13, 14, 62
   invention/design (distinction) 80
   inventor 7, 13
   knowledge based 74
   marketing 6
   materials 9
   paramount 8
   plethora 8
   product 2, 15
   realization 7
   recycling 8
   secondary 13, 81
   sell 4, 7
   sources 6, 78
   technological 6
Impact of technologies on markets Table 1, 64

Indicative tests 165
Industrial designs 80
    liaison 49
    liaison groups 49
    research associations 50
Information exchange 60
Innovation
    agent of change 7
    analysis 1
    applications 15
    aptitudes 3
    artist 8, 9
    centres 55, 58
    collective 7
    creativity 2, 4, 5, 13, 14
    definition 4, 6, 8, 9, 14
    design 36
    design led 100
    and entrepreneurship 10
    evolutionary 2
    experience 3
    experimental led 101
    guidelines Fig. 9, 66
    how, when, where 2, 8, 9, 11
    idea 2, 6, 8
    idea led 71
    inception 13
    invention 6, 7, 23
    logicality 1, 2, 4
    management 14, 67
    manufacturing 9, 10, 63
    materials 3, 14
    modus operandi 2
    objective 9, 13
    paired event motivated 72, 73
    phase summaries 14–18
    philosophy 1, 13
    practice of 150
    practitioners 6
    principles 10
    process 1, 2, 4, 7–11, 14
    product 1, 3, 6
    prototype 4
    secondary 13, 15, 80, 97
    shift 14, 31, 88, 100
    single event motivated 70–72
    sources 8, 10, 70–74
    strategies Table 2, 68
    targets 9, 11
    traditional view Fig. 1, 10
    triple event motivated 73, 74
    unit operations 19
    chain 13, 15
    performance 14

Innovation Project Model Fig. 3, 1, 6, 11, 12–14, 18
Innovator 3, 5
Innovator team 3, 51
    environment 46
    marketeer interaction 79, 112
    professional 53, 54
    role definition 54
Innovator's environment 46
Inorganic fibre mat Fig. 13, 103
Inorganic fibre stack block Fig. 14, 106
Inorganic materials 65
Interactions
    coherence 13
    events 11
    idea, experiment, invention 21
    innovation 1, 8
    invention 7
    market/technology 11
    modus operandi 13
    personal 13
    transverse/lateral 11, 12
Interactive model Figs. 2 and 10, 10, 11, 70, 71
Invention
    applications research 13
    commercial value 15
    definition 23, 80
    design 9
    development 9
    experiment 7, 13
    exploitation 16
    feasibility 15
    how, when, where 8
    idea 4, 9, 13, 14
    idea and experiment 15
    information recycling 13
    innovation 6, 7
    opportunity 9
    patenting 23
    process 91
    product 15, 23, 24
    secondary 13, 23, 36, 80
    surprise 22, 61
Inventive 2, 14, 79
Inventor 2–5, 7, 9, 11, 13, 14, 16, 63
    definition 63
    entrepreneur (isolation) 3, 5, 7, 16, 17, 51, 55
    improvisation 23
    personal computers 59
    sponsorship 58
Irex 59
Irrelevant investigations 65

Joint evaluation 124

Karmen Vortex Trails 49
Know-how 13, 24, 139, 140

Laboratory process 90
Laboratory process unit design 37, 127
Lay public 6, 8
Leasing 13, 17
Legislation 17, 129
Licensing 17, 138, 143
Logical
  analysis 1, 2, 14
  development 8, 14
  events pattern 1
  idea translation 14
  innovation 1, 2, 15, 17
  model framework 1, 15
  organization 14
  origins 4
Logicality and creativity 2, 4, 8, 13, 14
Loss leaders 28

Management
  business 17
  effective 14
  experience 1, 3
  innovation 67
  project 14
  strategy 47
  understanding 11
Manufactured products 6
Manufacturing
  feasibility 27
  industry 9, 63
  process unit operations 90, 91
Market 10, 16
  appraisal 6
  assessment 62
  collaboration 98
  criteria 16
  evaluation 16
  fit 139
  knowledge 27
  needs 10, 11
  places 3
  potential 60, 61
  selling 10
  size 26
  surveys 27, 31
Market pull model Fig. 1, 10, 16
Market research 15, 16, 17, 31, 38, 98
Marketable product 8
Marketeers 27, 63, 115
  innovator partnership 30

Marketing 2, 3, 6, 7, 17
Marketing organization 27
Marketing teams 52
Materials, inorganic 65
Matrix properties 105
Maximizing demand 97
Methodology 4, 6
Microelectronics 65
Modem systems 59
Monolith 113
Monolith support Figs. 16, 17, 18 and 19, 114, 116, 117

National Physical Laboratory 49
Negotiation 45, 146, 169
New knowledge 10, 65
New materials 11, 14, 65
New ventures
  business 13
  environment 7
  project 1–3, 5, 8, 9, 14, 18
  research 2, 22
  research group 52
  team 8
'Not invented here' syndrome 149
Novelty 6, 53, 60

Objectives
  definition 9
  design 9
  events 19
  product 6, 8
  targets 9
Opportunities
  business 4–6
  creative 5
Option rights 44, 96, 170, 148

Parker, R.C. 48, 66–68
Particulate filter 128
Patent
  Application for 93, 143
  applications research 120
  claim 156
  investment risk 23
  maintenance fees 144
  property 13, 16
  review 43
  rights 4, 14, 17, 147
  search 156
Patent Agent 26, 95, 120, 128
  fees 144
Patent Office 26
  application fees 144

publication 80
Patenting 2, 16, 23, 25, 139, 143
　added value 23
　strategy 155
Pattern
　events 1, 9, 11–13
　phases 13
Performance criteria 36
Personal computers 59
Phases 1, 11, 12
　applications 15, 16
　development/design 16, 17
　exploitation 17, 18
　feasibility 15
　innovation 14, 15
Philosophy 1, 2, 4, 6, 11, 63, 111
Picasso 9
Polystyrene composite foam Fig. 12, 80–88, 85
Polyurethane composite foam 150
Polyurethane composite rigid board Fig. 28, 170
Polyurethane flexible composite foam Fig. 23, 160
Post-impregnation treatments 152
Presentation 17, 45, 136, 145, 146, 149
Price structure 41
Priority exploitation 122
Prior-art 140
Process
　evaluation 90
　invention 91
　scale up 23
　secondary invention 24
　specification 125
　targets 63
Processing 6, 16
Product 1, 3, 4, 5, 6, 14, 16, 17
　aesthetic 36
　applications 2, 5, 15, 97
　base 13, 15, 16
　characterization 103
　cost 40, 98
　cost estimates 26, 40
　customers 7
　data 99
　data sheet 17, 146
　definition 125
　development/design 16
　end 1, 8
　evaluation 16, 88
　exploitation 2, 9
　feasibility 14
　field of use 25

finished 8
full scale 16
ideas 5–7, 74
innovation 1, 4, 6, 8
laboratory scale 16
limitations 28, 37
manufacturing 6, 9, 10, 14, 17
marketing 2, 8, 17
new materials 11
options 9
phases 14
price 38
prototype 4, 6, 16, 17
quality control 16, 118
resources 13
sale 4
secondary 15, 16, 80, 97, 100
specification 125
target 8, 9
targets (innovation) 63
test criteria 16
utilization 39
Product innovation phases 14
Production 4, 17
　capacity 126
　commercial 15, 16
　full scale 17
　in house 17
　manufacturing industry 6
　method 99
　plant 13, 14
　prototype 16
Professional 53, 54
Project 1
　adoption 45
　analysis 1
　applications 16
　assessment 34
　capitalizing 17
　case histories 1, 2, 9, 14
　champion 56–58
　change 7
　collaborations 7
　communications 7
　confidence building 27
　data summary 41
　development 16
　diversity 6
　evaluation 27
　experience 1, 3
　exploitation 15, 17
　feasibility 7, 15
　ideas 8, 74
　innovation chain 13

Project 1 (cont.)
  innovation model 1, 6, 11–13
    concept 11, 170
  innovative 8
  interactivity 13, 14
  inventive element 2
  management 14
  manufacturing 3
  new ventures 1, 5, 8, 9, 11, 14, 18
  objective 9
  presentation 17
  research team 7
  status 15
  target 1, 9
  team 8, 9
  team composition 54
  value enhancement 23
  variety 14
Promoting innovation 56, 63
Properties 15, 101
Prototype 1, 4, 61, 124
  design 34
  development 34
  evaluation 39, 58, 66, 160
  financial grants 58
  product 61
  specification 36

Quality control 16, 17, 30, 37, 99, 106, 118

Registered designs 145
Regulatory bodies 99
Reiterative process 13
Research 6
  applications 13, 15, 16
  department 52
  economics 16, 40
  experience 2
  institutions 7
  manufacturing 4
  marketing 16, 17, 92
  new ventures 3–6, 10, 14, 15
  project team 3, 7, 16
  samples 98
  targets 22
  teams 52
Resins 108
Resource
  allocation 66
  time limit 42
Rigid thermal insulation board 104
Role definition 54
Rothwell, R. 10, 11, 13, 61, 65, 101, 118
Royalties and fees 147

Saffil fibre (*see* Alumina) 104
Sale
  know-how 4, 13, 14, 17
  patent rights 4, 13, 14, 17
  product 10
Sample 123
Scale-up 23, 26, 36, 91, 158
Scientist 4, 63
Screening process 60
Secondary
  applications 16
  ideas 13, 80
  innovation 15, 81–87
    case history 82–88
  innovation model Fig. 11, 81
  invention 13, 24, 25, 30, 62, 80
  products 15, 16, 31, 80, 100, 126
Semi-technical plant 125
Sheet moulding 110
Silicate layer minerals 153
Simulation techniques 123
Small Firms' Advisory Units 144
Small-scale business support 55
Something up-front 147, 169
Sponsorship 58
Stack block furnace lining 105
Stack bonded fibre panel 105
State of the art 11, 56, 157
Statutory bodies, compliance 40
Statutory tests 36
Stress 48
Structural properties 105
Students 4
Students' creative potential 63
Support for innovation 59
surprise 61, 143

Tall towers, innovative design 48
Targets
  applications 28, 97, 98, 101, 122, 124, 159
  commercial 93
  innovation 63
  invention 77
  product 9, 33
  research 23, 31
  secondary 31
Team
  composition 54
  innovation 3
  innovator 3, 51, 55
    role 55
    travel 56
  new ventures 8, 9

project 7
research 3, 16
Technical Service 38
Technological
  analogy 9
  aspects 2
  capability 10
  change 7
  cost advantage 136
  exploitation 5
  knowledge 6
  performance 17
  state of the art 11
Technology Push Model Fig. 1, 10
Technology impact on industry Table 1, 64
Techno-commercial fit 30
Techno-commercial status 148
Terms, formalized 170
Test
  categories 39
  criteria 32
  manufacturing criteria 32
Testing 30, 52
Tests
  advisory 39, 40
  indicative 39
  mandatory 39, 40
  statutory 36
  user 39, 40
Time allocation 67
Time scale 59
Tolerances 99

Trade Associations 51, 57, 100
Transverse
  coupling 46
  shift 12, 46, 80
Truth 90

UK Fire Services 151
UK Patent Office Publications 145
UMIST 116
Unit cost 89
University Research 49
  teams 49
Upholstery-sample fire tests Figs. 25, 26 and 27, 162–166
USA
  Environmental Protection Agency 113, 128
  Government 129
  patent application 169
  truck and diesel Market 129
User
  collaboration 37
  specifications 32

Value enhancement 23
Venn diagram 12, 70
Venture capital 147
Visual aids 146

Waddington, C.H. 150
Wall cavity insulation 75
Weighing up opposition 156
Wind strakes (tall stacks) Fig. 8, 50